100
VIGNETTES

Happy Reading!

Jerry Hirsch

100 VIGNETTES

THE BOX *of* CHOCOLATES
APPROACH *to* EVERYDAY LIFE

JERRY HIRSCH

© 2012 by Jerry Hirsch. All rights reserved.

WinePress Publishing (PO Box 428, Enumclaw, WA 98022) functions only as book publisher. As such, the ultimate design, content, editorial accuracy, and views expressed or implied in this work are those of the author.

No part of this publication may be reproduced, stored in a retrieval system, or transmitted in any way by any means—electronic, mechanical, photocopy, recording, or otherwise—without the prior permission of the copyright holder, except as provided by USA copyright law.

The author of this book has waived a portion of the publisher's recommended professional editing services. As such, any related errors found in this finished product are not the responsibility of the publisher.

ISBN 13: 978-1-4141-2315-8
ISBN 10: 1-4141-2315-9
Library of Congress Catalog Card Number: 2012903983

Contents

Foreword ix

1. On the Bus 1
2. Doing the Grind 3
3. Driving Habits in BC 6
4. Do You Have a "Bucket List"? 9
5. Dealing with Change 12
6. Housecleaning Techniques 14
7. Shopping 16
8. Buns and Bread Don't Just Happen to Be on the Grocery Store Shelf 19
9. Hand Washing 21
10. A Brief History of Flying 23
11. Top Inventions—Only in Canada, Eh! 25
12. History's Most Brilliant Inventions 28
13. The Red Lantern 31
14. Restorative Justice, Forgiveness and Compassion 35
15. Should We Attach an Asterisk to Your Performance? 38
16. Can You Say "Merry Christmas"? 41
17. I Learned Everything I Need To Know in Kindergarten, From My Grandchildren and the Easter Bunny 44
18. Senior Cycling Adventures 46
19. Some Jokes and Tongue-in-cheek Comments 49

20. Airbus 380, the Boeing 747, and Other Heavier than Air Birds 52
21. Mother Nature's Intricate Details 55
22. The Human Body Is a Wonderful Thing. 58
23. English Is a Crazy Language 61
24. The Joy of Mathematics—An Oxymoron? 63
25. Technologies that Changed the World. 66
26. Everyday English and Interesting and Conflicting Expressions 69
27. Things I Remember as a Kid that Are Now Obsolete 71
28. Are Computers Male or Female? 73
29. Some Comic Strips Are Hilarious; Others Not So Much....75
30. Watch This!. 78
31. What Time Will It Be When We Get to Toronto? 81
32. Keeping Time 83
33. My List of "Most Admired People" 86
34. Is Anybody Hungry? 89
35. Snoring vs. Leprosy..............................92
36. We've Become Our Parents 95
37. Worship Music in Church. 98
38. What Is a Split Infinitive, and Why Would Anyone Want to Dangle a Participle? 101
39. Airbags and Other Interesting Chemical Reactions in Automobiles 103
40. Is that You Glowing in the Dark? 106
41. From Smoke Signals to Cell Phones to Twitter to....? 109
42. That's the High Price of Being Canadian, Eh?. 112
43. Ten Interesting Cities I Have Visited. 115
44. Buy Land. They Ain't Making Any More of the Stuff 118
45. Urban Farming—Look No Further than Your Own Backyard 121
46. Flip-flops. 123
47. Who Decided that Doctors Should Have Poor Handwriting? 126
48. Do You Know as Much About Canada as Americans Know About the U.S.? 129

49. You Know You're Getting Old When … 132
50. Important and Light Weight Philosophical Questions. 134
51. Living in the Far North or Other "Isolated" Places 137
52. Social Media and Computers Are Changing the
    Way We Live. 140
53. Big Brother Is Watching 143
54. Pure Science vs. Applied Science vs. Junk Science 145
55. Unions Aren't What They Used to Be. 148
56. Snail Mail Isn't Dead Yet, But…. 151
57. The Organic Movement—Is It Just Another Form of
    Activism?. 153
58. Why Is It So Hard to Export or Implement Democracy?. . .155
59. Why Do We Have Fan(atic)s? 158
60. Grandchildren Are More Fun than Children! 161
61. The Space Shuttle and Our Need for Galactic Adventures. .163
62. Some of My Favourite Quotations. 166
63. Sports—No More Dynasties, But More Greed 169
64. We Live in a Multi-Faith Society. 172
65. What Are You Afraid of? 175
66. Prayer—Why and How Do We Pray? 178
67. Millions, Billions and Parts Per Quadrillion 181
68. Cross Country Odyssey 2006 184
69. The Resurrection of Jesus 189
70. Minor Political Parties in Canada 191
71. Signs of the Apocalypse. 194
72. Only 56 Years Ago … 197
73. Winter in the Lower Mainland vs. the Rest of Canada 200
74. Everyone is Created Equal, But Some Are More
    Equal than Others. 204
75. Advertising 207
76. TOMS Shoes, One for One, and Mossy Foot 210
77. Why Can't Europeans Be More Like Us? 213
78. Here's a Hot Topic 216
79. The "Machinery of Government" and Other Things
    You May Not Have Heard About 219
80. Is the United Nations Effective? .222

81. Country Music 225
82. Sweet Hummingbirds 228
83. Conspiracy Theories and Theorists 231
84. BC—Why Would You Live or Vacation Anywhere Else? ... 235
85. Adjectives, Adverbs and Beautiful Churches 238
86. Quebec Wants a Divorce with Bedroom Privileges 241
87. "Check Your Value System at the Door" 244
88. Tantalum 246
89. No One Likes Food Additives, but Everyone Wants to Use Them 248
90. MRSA—The Four Scariest Letters in the Alphabet 251
91. Reforming the Senate 254
92. Plastic Money 257
93. Will that Be Credit or Overdraft? 259
94. Dirty Jobs and Hazardous Jobs 262
95. Painting Pictures of the Big Apple 265
96. Snowbirds 269
97. Wonders of the World 272
98. Canada's Wonders 276
99. The Ultimate Trivia Quiz 280
100. And Now for Something Completely Different—Harry and the Leeches 284

Foreword

WHAT DOES ONE do when you are partially retired? Unless you are a Floyd Landis—Tour de France type of guy you can't cycle at night or in the rain and snow; there are only so many games of Penny Rummy that you can play; there is only so much strength you have to do things with the grandchildren; and while reading is a rewarding occupation it takes at least a modicum of concentration. So I turned to pecking at the keyboard.

Many of the "vignettes" are based on my own experiences while some information has been gleaned or summarized from Wikipedia, but I often have added my own thoughts and views in these cases as well. Other ideas have been taken from various newspaper columns and news articles, while other items arose just by observing life each day. In some cases, ideas have been bouncing around in my cranium and somehow found their way on to the printed page.

Many of the musings are Canadian oriented, so the reader from another country will either learn a bit about Canada or just move on and open another vignette.

Everyone has an opinion, even on issues where they are not particularly knowledgeable, and I'm no exception, as you will see, on the amateur political commentaries or the essays on financial issues. I have a little more expertise in science and social questions, but then I'm often wrong there as well.

VIGNETTE 1

On the Bus

WORKING IN DOWNTOWN Vancouver, BC while living in the suburbs, as I did for many years, has some distinct advantages. For example, besides having a myriad of choices if you want to go out for lunch or go for a quick noon time walk, I had lots of time to notice people and observe their habits as I rode to and from work each day on the city bus. People come in all shapes and sizes and mannerisms, and watching them can be almost as interesting as work.

There's the portly old gentleman tipping the scales at 295 pounds (134 kilos if you really need to know) who takes at least 1½ seats and really should pay extra fare. There's the sweet little old lady going home at rush hour with three or four overflowing shopping bags when the bus is so jammed full that vacant seats are hard to find, lamenting to her equally burdened friend that people aren't very kind and, "Isn't it a shame that chivalry is dead?"

There are the bus drivers who try to make the trip more pleasurable by turning their buses into saunas. And some of the passengers must be participating in deodorant commercials since they refuse to open the windows in the summer.

There are the fellows who insist on taking home on the bus a curtain rod or a nine foot tail pipe for a half ton or Mack truck. These characters show infinite patience and clumsiness in extracting their treasures from between the seats, while bumping other passengers on their shins or shoulders just before getting their prize item stuck in the back door.

Then there are the passengers who sleep. Some of them must have excellent internal clocks since they sleep very soundly until their exact stop is reached, at which point they suddenly wake up, rearrange themselves and collect their packages or briefcases, and charge to the exit door. Others are less controlled and gradually rest their head on the passenger's shoulder next to them. A person with an imaginative camera could make a fortune.

There are the folks with the penetrating or foghorn voices, who share their intimate activities and secrets with their friends. I recall two ladies who regularly put their voices and mouths in high gear and shared all kinds of personal stuff from the moment they got on the bus until they got off. The ride from Richmond to downtown Vancouver took about 55 minutes, so I could have made a lot of money if I had been a writer for a syndicated soap opera program.

There are interesting characters that make everyone else look quite ordinary. One fellow must have been demonstrating the R-value of his hairy chest, since even in winter he wore his black cowboy shirt (he only had one, which he wore everyday as far as I could tell) open to his navel. Maybe it was something in the little black bag or pouch he always carried. The bag matched his black cowboy hat, black pants and black boots, and even the extra-long handlebar moustache.

Then there are harried mothers trying to shepherd their four preschool kids on a crowded bus. What would you do if a 15 month-old baby sticks her sucker onto your newspaper or into the hair of the crusty businessman in front of you? There are passengers who read their newspapers every day but never quite master the technique of folding a newspaper in a confined space. And perhaps you have seen passengers who never bother to buy a paper or a book but are fascinated by what the person next to them is reading.

And then there are passengers who write notes like this on the bus.

VIGNETTE 2

Doing the Grind

IN MY ADVANCED age and avoirdupois I have periodically taken up the challenge of the Grouse Grind, which is a three kilometer (1.8 mile) trail up the face of Grouse Mountain in North Vancouver, BC. The Grouse Grind is often called "Mother Nature's stair master," since its almost 2900 steps bring the adventurous hiker 853 meters (2700 feet) higher than the starting point, almost to the top of Grouse Mountain at 1130 meters (3700 feet) elevation.

This trail is very challenging and not for the average Sunday afternoon stroller (they should take the gondola up the mountain). The first time I tried the Grind several years ago it was with some other similarly fitness-challenged folks, and since we took our time (a lot of time) it seemed almost manageable. But I got a taste of what it could be like if one actually took this a bit more seriously.

I managed to do the Grind six times a few summers ago, and I learned something each time. The most important lesson was to go at my own pace. It's easy to be discouraged when you see guys and girls who seem to be on an escalator and hear them saying something about finishing in less than 47 minutes. I have come down from 90 minutes to a best time (so far) of 62 minutes, which isn't bad for an old guy, or so I keep telling myself. Instead of using a fast paced "escalator" pace, I learned to do what I call my "fullback double step"—like in football you have to always keep your legs moving, but yet not so fast that you stumble or wear yourself out in the first half of the trek. On the Grind one has to keep moving forward or else you tend to fall back on the high steps. And it is an excellent cardiovascular exercise.

I have learned that stopping is a big time user (surprise!), and for a while I managed to stop only at the quarter, half and three-quarter marks, and then only long enough to get a drink. With a bit more practice, I've found that not stopping at all is actually possible. It's a good idea to carry a water bottle, but I find it best to drink lots of water before starting out, and then while cooling down at the top. The purists may not approve, but I find that a walking stick helps—it gets ones arms and upper body involved and provides balance.

On one of my early trips I was so tired about 3/4 of the way up that when a mosquito tried to bite me I barely had enough strength to swat it away and not enough energy to actually do it any physical harm. It's rather funny, meeting people on the trail and greeting each other with a muffled breathless voice that says "I can't waste my very limited amount of energy by saying hi in a normal way". And then there's the age old cry of, "Are we there yet?" in greeting people who are coming down. Coming down is a whole other story and is generally not recommended, since it's very hard on the trail and your legs; taking the gondola down is easier and inexpensive.

Perhaps the Grind is comparable to golf. Maybe I can "hike my age" in minutes as I get older? I took a friend with me recently for his first trip, and he went through the same things as I did on earlier forays. After the first two hundred meters he discarded his baseball cap, and after 10 minutes he took off his jacket. After 30 minutes he was asking, "Why are we doing this?", and at the top he was wishing that he had brought along a spare (dry) T-shirt and contemplating if we should still be friends. By the way, if you're really keen, you can obtain a radio frequency chip so that you can swipe your card at the bottom and the top and be entered into their data base of ultra-dedicated Grinders. Last I heard the record was about 25 minutes for men and 31 minutes for the ladies.

Doing the Grouse Grind is quite different from the annual 10K Vancouver Sun Run in that the Grind seems more physically demanding, although it probably isn't. Maybe it's because I used to do the Sun Run in just under an hour. I remember the Sun Run one year where at 5K I

Doing the Grind

thought, "Well, at 6K I'll turn on the after-burners to get a good time." When I got to 6K, I thought, "Well, maybe at 8 K". Then I got to 8K only to find out that the after-burners had long since fallen off and I had to muster on as best I could. Anyway, I just kept telling myself that all this exercise is good for me.

VIGNETTE 3

Driving Habits in BC

APPARENTLY "STOP" REALLY means, "maybe slow down" to many drivers in the BC Lower Mainland. With this in mind, I have an idea that I would like to pass along to the police and those who set our traffic laws—let's change almost all of the STOP signs to YIELD signs. I'm serious about this—it makes sense because the typical driver almost never stops at a STOP sign (the only exceptions seem to be drivers with a driving instructor beside them, drivers over 70 years old, or when a police car happens to be nearby). Sometimes drivers will actually *slow* down at a STOP sign, and if the traffic is going right past their nose at 75 KPH (in a 50 zone, mind you), the driver might actually stop. But the majority of time they cruise into the intersection, and if there isn't a vehicle within, say, 20 or 30 feet, they just keep going. Another option would be to install speed bumps on more streets so that most people would actually slow down. Right hand turns at an alleged STOP sign seem to be a particular invitation to just keep going. I've seen many drivers not even slow down while making a right hand turn. Rather than invite total disregard for the law, let's change those STOP signs to YIELD. This would make sense in many residential streets anyway, even if drivers actually obeyed the signs.

At the opposite end of the awareness spectrum are the drivers who have never figured out how to use an access lane. These folks sometimes actually stop, right at the start of the access lane. And they wait, and wait, and wait a bit more, until there isn't a car within 2 or 3 kilometers, or the foreseeable future, whichever comes last. It's too bad for the rest of us that they missed driving school the day that merging into the traffic flow was covered.

And why don't we raise the speed limit by 20 or 30 KPH while we are at it? I mean it's getting dangerous having so many drivers rush up behind you, and if you're lucky, past you, at more than 85 or 90 KPH while you're doing 70 in a 60 zone. It seems that no one really believes the posted limits, although the presence of a police car has a (temporary) magical effect in slowing down the traffic. Parked vans used to have the same temporary effect when photo radar was being used in the Lower Mainland of British Columbia. Many good drivers would argue that speed limits should be raised on most major streets in any case. One more thing—perhaps we could even pass a law that a vehicle must be in a lane for at least 15 milliseconds before switching to another lane.

One of my favorites is the guy in the gravel truck or the 28-wheeler semi who thinks he's still driving his half-ton truck. Weaving from lane to lane, tailgating like they can stop on a dime, setting their cruise control at 25 kph above the speed limit; why it's enough to make you vow to buy a monthly transit pass. The obvious exception to this of course are those new truck drivers taking semi-trailer truck driving lessons and cautiously going down the street at 18 KPH, but that's only temporary and they will soon be speeding when they find 10th gear.

This probably isn't the place to note this, but BC drivers are generally better or at least more careful than those in some countries. Italian and French drivers are fearless and aggressive with much blowing of horns. I found that there weren't many cyclists in Madrid or Barcelona or Paris; I suspect that they were all either maimed or discouraged by the crazy drivers. It was quite interesting to see the 8 lanes of traffic tearing around the Arc d'Triomphe in Paris round-about with 7 entrances/exits and space for at least 6 lanes, but no lane markings—great courage or at least experience and good peripheral vision was necessary for all drivers. I can see why Chevy Chase in European Vacation went round and round here for many hours without finding a way to get to an outside lane and an exit.

They say that every day brings its own surprises. Perhaps one day we will find some drivers who respect the law and each other.

VIGNETTE 4

Do You Have a "Bucket List"?

HOW WOULD HAVING a bucket list change your life? A popular movie a few years ago made popular the idea of a "bucket list". Two interesting characters from totally different backgrounds find themselves in hospital facing the inevitability of death from terminal cancer. One of them remembers that many years before in college a professor gave an assignment to fill out a bucket list of the things they might want to do in life before "kicking the bucket." Facing an uncertain but limited future, this fellow starts writing his list of things to do before he dies. His new friend sees the list and adds some other things that he always wanted to do. The activities on either list include: witness something truly majestic, help a complete stranger, laugh till I cry, kiss the most beautiful girl in the world, sky dive, and visit Stonehenge.

Rather than slowly die under medical supervision, the two compatriots decide to check themselves out of the hospital. They embark on a journey of friendship, discoveries and redemption by completing a variety of adventurous and unusual activities that they had identified on their bucket list. They leave on an around-the-world vacation, doing things like race car driving, skydiving, climbing the Pyramids, and going on a lion safari in Africa. They discuss and experience a rare coffee and its unusual taste. Along the way they become close and confide in each other about faith and family.

There are several issues here. One is obvious; have you developed a "bucket list" of things that you would like to accomplish before you pass on? This can be a good thing; it might inject some excitement and adventure into our ordinary lives. But more significantly, preparing such a list, and making some serious effort to follow through, is likely to lead us to examine our lives, to move beyond just doing something ordinary or interesting, to doing a better job each day of relating to those we love and helping those less fortunate than ourselves. I'm reminded of the late Emory Barnes who decided to spend several nights on the streets of Vancouver to better understand how the homeless cope.

Here's an idea for something to put on your bucket list: take an adrenalin filled view off the "Edge Walk" at the top of Toronto's CN Tower. You would be tethered to an overhead guide rail and can tiptoe to the very edge of the platform to have a stunning view of the Toronto skyline 356 meters (it sounds more challenging if you say 1167 feet) below you.

So perhaps we, whether we are strong and healthy or not, should develop two bucket lists: one that includes random acts of kindness that are likely to make us a better person, and another one that injects excitement into our ordinary lives. Why not give this some thought? Here's a start: My "A List" includes: visit someone in prison, pay the bill of the person behind you in a Starbucks or Tim Horton's coffee shop line, send an unexpected greeting card to family members, become more patient, send a note of encouragement to my Member of Parliament, attend a city Council meeting, take more walks with my wife, read at least one book a month, take more mountain hikes with my neighbor, and volunteer with Prison Alpha.

My "B List" includes visit Europe, hike Mt. Kilimanjaro, spend a month in Ireland or some other exotic location, do the Grouse Grind at least twice a year, cycle at least 2,000 km next year, and go on a cruise. And then there is the ultimate Bucket List—enjoy eternal life with God, which will be light years better than any experience on earth. What does your list look like?

Here are some ideas that may stimulate your thinking: My friend Willi regularly sends me informative selections from the Internet, such as the following list that I call "Twenty-five Pieces of Advice" about

Do You Have a "Bucket List"?

making life enjoyable. (1) Walk 10 to 30 minutes each day and smile while you walk. (2) Sit in silence for at least 10 minutes each day (lock yourself in if necessary). (3) Listen to good music every day; it's food for the spirit. (4) When you wake up each morning, set at least one goal for the day. (5) Live with the 3 E's—Energy; Enthusiasm; Empathy. (6) Play more games than last year. (7) Read more books than last year. (8) Look at the sky at least once each day and realize what a magnificent world in which you live. (9) Dream more while you are awake. (10) Eat more natural foods and less processed foods. (11) Eat berries and nuts, and drink green tea and lots of water. (12) Try each day to make at least three people smile. (13) Reduce disorder in your home, your car, and your desk. (14) Don't waste precious time gossiping or regretting the past, or worrying about things beyond your control. (15) Realize life is a school, and you are here to learn. (16) Don't miss any opportunity to hug someone you appreciate. (17) Don't take yourself too seriously—no one else does! (18) Don't compare yourself with others. (19) Remember that you don't always have control over what happens to you, but you do have control over what you do with the experience. (20) Learn something new every day. (21) No matter how you feel, get up, get dressed, and be there. (22) Call your family regularly, and send them messages saying that you are thinking of them. (23) Remember you are too blessed to be stressed. (24) Enjoy life—you have only one opportunity to make the best of it. (25) Send this list to someone you care for.

VIGNETTE 5
Dealing with Change

WHY ARE COINS made of metal? Answer: because change is hard. The varied and sometimes emotional opposition to holding the 2010 Winters Olympics in Vancouver/Whistler reminded me that one can always find people opposed to a new idea or a new thing. These people must feel it's easier to criticize than it is to stick your neck out and support a bold new idea.

I'm no great fan of the Olympics, since they often promote super-nationalism, and since the idea of the modern Olympic movement was based on class distinction when only the wealthy could afford to be "amateurs". Yes, there are some cost overruns, yes they will cause some substantial disruptions, and yes the "Olympic family" is still pretty exclusive, but on the other hand they also celebrate the human spirit and achievement. However, the Olympics are not just a two week party for the wealthy. There were hundreds if not thousands of visitors and officials that came to Vancouver and Whistler since the games were first awarded there, plus there were many athletic teams who came to compete and train. And there is an expectation that thousands and thousands more visitors will visit BC months and years after the Olympics have ended. These visitors pay the salaries of a lot of taxi drivers and cleaning staff and waiters and bellhops and airport staff.

We could dwell on the disruption and the cost and the unfair perks (e.g. tickets given to the politicians). But let's also consider some of the long term advantages such as the new facilities like the Olympic curling venue that was modified to include a gym and library. The Olympic speed skating oval was converted into a facility that many in suburban Richmond and the surrounding area will use for years to come. And the list goes on. Then there is the legacy of the Canada Line to improve city transit, and the Convention Center, as well as the upgraded Sea to Sky highway that will save lives and make a day trip from Vancouver to Whistler easier and safer for all of us in the future.

The 1960s were full of change, and many people were there to vociferously oppose new ideas and plans. Remember the changes on the federal scene? Remember how some Canadian politicians almost lit their hair on fire in the debate over creating a new Canadian flag? I'm old enough to remember the threats that doctors would soon be leaving Canada en masse when parliament was passing legislation on universal access to health care. And Prime Minister Pearson and his government endured huge criticism in passing legislation establishing the Canada Pension Plan. Or how about the Feds implementing the metric system in Canada and how the naysayers said this would severely damage trade with the U.S.? Now, all of these supposedly hugely controversial issues are pretty much accepted and we haven't suffered the ignominies and disasters predicted by the doomsday naysayers who were afraid of change.

Why are people often opposed to new ideas? Sometimes, as in some of the cases above, there is self-serving political gamesmanship. I suppose that happens frequently in politics at all levels where it just wouldn't do to credit your opponent with a good idea that should be implemented. Perhaps some can't stand seeing others realize success and experience reward for having the courage to take a chance. Sometimes, people just like things the way they are, because life is good, so I suspect that change often removes their security blanket and forces them to expose their weaknesses or to actually take risks. Perhaps some people are just plain old contrarians. Sometimes this is a good thing in moderation, sometimes caution is a good thing, but in many cases we would be much better off if more people would instead say, "Well, that's a good idea that should be implemented, and we could make it even better if we made this change and looked at this factor and, then, well, let's do it."

VIGNETTE 6

Housecleaning Techniques

SOMETIMES, WHEN I have an escape route well planned, I tell my wife that I'm better at cleaning the house than she is. After all, I say, tongue firmly planted in cheek, on those rare occasions when I need to do the cleaning all by myself, I can clean the kitchen in 10 minutes flat whereas she normally takes an hour and a half. If you watch carefully, you can see pigs flying past the window. It turns out of course that I miss a few spots, but it seems unfair to get into such picky details. Cleaning the house has to be one of my top ten unfavorite things to do, right up there with going to the dentist for a root canal or experiencing food poisoning.

You may recall descriptions in George Orwell's "1984" where the political masters put hairs or small items in certain places to be able to know if such places were being examined or violated. I sometimes wonder if that's what my wife does to check to see if I'm doing a good job of my share of the cleaning. Actually she doesn't really need to do this—it defies the laws of probability, but she can always find the one spot that I missed when dusting.

I have a lot of respect for people who clean houses for a living, but it's not a career path that I would look forward to. We had a house cleaner for a while, but when we spent more time cleaning before the cleaners came so they wouldn't think we were Neanderthals, we decided it wasn't worth it.

Housecleaning Techniques

Some people are fastidious about keeping their house clean (they hand you white gloves and hospital-like slippers at the door), while others are quite content to have their home in such a mess that it looks like Napoleon's army had just passed through. Some perfectly nice folks just don't care about this house-cleaning thing, and this approach has some advantages. For example, think of how much less wear and tear there is on the carpets if clothes are tossed all over the floor deep enough to temporarily lose a toddler. And how about the savings on laundry detergent? The vacuum cleaner doesn't need to be fixed or replaced, and the dishwasher works better as a cupboard anyway. Why cook dinner and dirty a lot of dishes when that pizza from last week, still on the kitchen counter, still looks good? But the Health Department should get involved in some homes, where the mess is not only disgusting but may actually be a health hazard due to mould and bacteria being given golden opportunities to succeed.

While we are speaking tongue in cheek, who decided that smooth was good and wrinkled clothes were bad? In a democracy we should all have more input into important decisions like this. I wouldn't be surprised if some manufacturing think tank was sitting around one day brain storming ideas, and someone came up with the bright idea of inventing the iron, and soon they formed two companies—one was a company to manufacture irons and one was an ad company to sell the idea that wrinkles were not cool and—hey, we have something to fix that.

VIGNETTE 7
Shopping

"JACK SPRATT COULD eat no fat; his wife could eat no lean. So just between the two of them they licked the platter clean." This children's nursery rhyme summarizes the different approach to shopping by men and women fairly accurately.

When we get the weekend paper or the local free papers, I look at the news and sports parts of the paper, and of course the comics. But the paper isn't worth getting, my wife says, if it doesn't have a zillion pages of ads and flyers. We both studiously go through our respective sections and the world keeps turning as it should.

I've heard it said that men go hunting while ladies go shopping. A successful "shopping trip" for me is one where I go into a store, even a huge box store or department store, find what I need, and am back in the car within 10 minutes. (For my wife, even if she starts out with that intention when with me, inevitably several things will distract her and the 10 minute trip turns out to be 55 minutes.) For most ladies, a successful shopping trip is one where they visit almost every dress store in the mall (and there are a lot of them!), and then they check things out at two or three other malls. It doesn't matter that they finally buy the outfit that they saw at the very first store (although this may be against the rules). And this doesn't just apply to shopping for dresses or other items of ladies clothing. I think that the ladies call this "retail therapy". With the possible exception of groceries, I think it violates some unwritten code

Shopping

for ladies to buy the first or second thing that they look at in a store and like it. Here's a conundrum: I go shopping for clothes once or twice a year and I have more clothes than I need while my sweetie goes clothes shopping several times a month and often says, "I don't have a thing to wear." I'm just speculating, but I think she means that she doesn't have anything new to wear. But I guess it should be mentioned that I usually look like Johnny Hayseed while she looks like a Sears catalogue model.

Some social scientist should write a PhD thesis about shopping in Craft Stores or Ladies Wear Stores. I'm not brave enough to venture into them very often (and never alone), but the most obvious thing that you see are the bored husbands wandering the aisles, just like the Kingston Trio song about guy who could never get off the train and go home after they raised the fare while he was riding the Boston MTA. All Ladies Wear stores and Craft stores could increase their profit margin if they just had a small coffee bar with a few magazines and newspapers (or perhaps books like War and Peace) for the poor guy who is waiting for his spouse to find that elusive material for a new craft or try on most of the dresses on rack number 17. Why is it that you never see bored wives walking the aisles of the auto parts store? A possible explanation is that they are down the street busy buying dresses. I guess the female side of the species is just a little more savvy when it comes to shopping.

The only stores where men and women seem to have substantial common and overlapping interests are grocery stores and the modern sports equipment super stores. We all like food, and more and more people (and not just young people) are into buying sports equipment and clothing. The Internet provides excellent alternative shopping opportunities for those folks who dislike shopping as we have known it to this point, and once all of the security concerns are satisfactorily wrestled down I suspect that most of us, or at least the men, will do much of our shopping this way.

Some people say that it's hard to buy gifts for men. Tools are always a good idea even if the guy is like me with no discernable practical skills. But here's another idea: an interesting book or magazine, or a specially designed soft toilet seat. A man's home is not his castle; instead his bathroom is his castle. This is the only place that a man can read the paper or a book without fearing constant interruption, at least if there's a solid lock on the door. But the problem is that toilet seats are hard, and after a prolonged (reading) session one tends to lose feeling in the legs. At least, that's what I heard. One time when men are forced to go shopping is when they need to buy something for their sweetie for Christmas. Many guys, at least those with a strong constitution, don't see anything wrong with leaving this to December 24th. Unfortunately, while I could compose a list of things to buy for men, I don't have any other ideas to help my fellow male shopping sufferers when they need to go shopping solo for their sweetie.

VIGNETTE 8

Buns and Bread Don't Just Happen to Be on the Grocery Store Shelf

ONE OF THE coolest things I ever did was to show two of my grandsons (then 6 and 8) that there were a couple of steps in the process before a bag of buns or a loaf of bread magically appears on a store shelf. I was raised on a farm, but they were city kids and weren't aware of all the things that have to happen before a product like bread is available for purchase and eating.

At the time we lived on a sizable lot with a small garden, about 60 square feet, divided by planks into 4 boxes. After digging-in the compost and hoeing the soil well, we tacked lines of string about 3 inches apart, (about 8 cm for the metric folks) and then hand planted kernels of wheat obtained from an Alberta farming relative. I carefully watered my 4 small "fields," and every time the boys came over we checked the progress of our crop as it gradually changed from small green shoots to nice tall plants with heads of wheat soon appearing, to green-golden stalks of wheat. It was agony as we waited for the water and the sun to do their magic, but eventually the stalks were at least 3 feet high and the kernels of wheat turned a golden yellow and became hard.

Finally, it was time for harvest. We took a pair of scissors and cut the heads off the wheat stalks and laid them on a large tarp. The boys used small 2 x 4 blocks of wood to do the "threshing" in separating the husks from the golden kernels of wheat. Then they took Nana's hair dryer and blew the chaff away, and scooped up the wheat kernels. We ended up with an almost full four liter ice cream bucket. I weighed the bucket and since I knew how much a bushel of wheat weighed and the area of my "field" I was able to calculate that the yield was 90 bushels per acre, which is at least two or three times the average yield on a typical prairie farm, but then they can't give each stalk of wheat the kind of personal care that we did.

Now came the next step. We took Nana's coffee grinder and ground the wheat kernels into flour (we ended up burning out the grinder near the end, but that's another story). Since the dough would have been too heavy just using the wheat flour, Nana added in some regular all-purpose flour and helped each of the grand children make buns. They eagerly watched the buns rise and then bake, and consumed them right out of the oven after adding slabs of butter. It was a great experience as they learned how buns and bread come about before these items get to the grocery store.

Actually, we learned another lesson—that in farming a great crop isn't guaranteed, due to drought or too much rain or some pestilence. The story above was year two of our project. The first year there were some old buildings torn down a few lots away, and we had a rat problem. Just as the stalks and kernels were a nice juicy green, the rats somehow pulled the wheat stalks over and ate the wheat, leaving behind little piles of husks. In about 3 days our crop was pretty much devastated. I "ploughed" over my little fields and vowed to "wait until next year". We didn't have a rat problem the next year, and our second try was successful.

VIGNETTE 9

Hand Washing

I RECENTLY ATTENDED a seminar on hand washing (I know, I know, I should get a life with more action) that I found quite interesting. The speaker said that, after using a washroom, one should wash their hands long enough to be able to sing "Happy Birthday" or "Twinkle Twinkle Little Star" at least once all the way through. Another interesting thought was that garbage cans in public washrooms should always be close to the door so that a person can open the exit door with paper towel in hand, and then deposit the used towel in the garbage can as they leave. The speaker went on to say that it is perfectly acceptable to throw the used paper towel on the floor by the door if the garbage can is too far away, and if enough people do this eventually management will get the idea that the garbage can is in the wrong place. By the way, washing your hands for an inordinately long time, like more than 3 minutes, isn't a good idea either since this dredges up subsurface bacteria.

    I tested some of these statements in our new Abbotsford city hospital when my wife was an inmate there for a few days. The visitors' washroom had a lovely garbage can next to the sink but too far from the door, so I moved it next to the door without impeding access or exiting. Later that day the garbage can was back beside the sink. I moved it back by the door but by next morning it was back by the sink, so I deposited my used paper towel on the floor by the door and I noticed later that someone else had done the same thing. But the next day the garbage can was still by the sink. My wife wasn't in the hospital long enough for me to see a change in the learned behavior of either the cleaning staff or their managers. I guess that hospital bureaucracy is something like government bureaucracy. A related and even more significant concern is that, according to surveys even doctors, and to a lesser extent nurses, don't wash their hands as often as they should in going from patient to patient in a hospital.

I saw a variation of this problem in a government laboratory. They had a nice big sign pinned to the exit door showing a pictorial of two hands holding a paper towel, with the statement "Keep Them Clean," implying that one should use a paper towel in opening the door. But they didn't get it quite right since the built-in garbage can was nowhere near the exit door!

I have two stories demonstrating that some people don't have an appreciation of how easily bacteria, including the disease-carrying kind, can spread by person to person contact. A friend was in the washroom of a large store and saw someone leave the stall, open the exit door and walk out. A few minutes later this unwashed customer was observed handling toys in the store, so my friend decided not to purchase anything at the store that day. It's enough to make one wish that the greeters in large stores provide everyone with neoprene gloves as they enter. Perhaps this is a job in which NHL Commissioner Gary Bettman could do well. My other story is even worse. A government inspector in another Canadian province was inspecting a large food manufacturing plant, and used their washroom just before lunch. In spite of large signs imploring staff to "Wash Your Hands Before Going Back On The Line," two employees were observed leaving without washing their hands. When challenged, their response was, "We're not going back on the line; we're going for lunch." The inspector quickly required some additions to the company sanitation training program.

VIGNETTE 10

A Brief History of Flying

IT SEEMS THAT everyone has an unhappy story to tell about an airline flight. I'm so old that I can remember some good things, like when Canada's two major airlines served real food ("would you like a steak or chicken dinner") and they actually used china plates and genuine silverware. I can also recall my first flight back in the days when airports had little kiosks that sold flight insurance. I guess the strategic thinkers at the airlines eventually decided that this didn't create a positive safety image.

I know that families with babies need to travel sometimes, but couldn't they use their mini-van? Have you ever been stuck in a middle seat, or even a window seat, with a baby crying non-stop next to you? Or how about having a two year-old almost deposit everything they have eaten for the past 34 hours on themselves and you? When this happened to me, the airline crew spread a pound of coffee over the entire area, but pretty much were indifferent to cleaning up the mess. But you may know the motto of Canada's national airline: "We're not happy till you're not happy." This is the same airline that was charged with misleading advertising, since they hid substantial extra costs in the fine print while advertising "discount fares". In a number of surveys, crying babies were identified as the highest irritant on airplanes, even higher than sitting beside someone with terrible body odour or the boring person who never stops talking or complaining about the shrimp-sized seats.

For long trips the airlines have devised a special type of torture—they march you past the recliner seats in business class before they jam you into a sardine-size seat in the economy section. And of course you know about the new cost-saving fad being imposed by airlines. "You have two bags, well that will be $57.50 extra. Dinner, well that will be $7.50 for the sandwich that was specially brought in from Hong Kong four days ago. Pillows and Blankets?!!—you have a neat sense of humor, ma'am!" I can't wait for the day when there will be a surcharge if you want the plane to land safely.

There has been some activity recently regarding "passenger rights"—now there's a concept. Airlines are not supposed to let passengers sit in a plane at the gate for more than two hours without offering food and water, for example. One of the most irritating things is when a flight is delayed and you are sitting in the terminal without being given accurate information. "Flight delayed 30 Minutes" is repeated several times, and finally one or two hours have gone by without any details—then it turns out the flight is cancelled. But passengers have responsibilities and idiosyncrasies as well. Some of the things folks try to bring on board boggle the mind, but fortunately, modern day security checks have taken the wind out of the sails of many of these folks. Some passengers get pretty obnoxious after imbibing a few pops, creating problems for both fellow passengers and airline staff.

Bad weather, particularly in Canada in winter, can throw a monkey wrench into the best laid travel plans. Murphy's Law predicts that weather-related flight delays and cancellations will happen at the busiest travel times, such as during the Christmas holiday season, At least one of Canada's airlines seems to try very hard to do a good job; the agents at the counter are actually friendly as is the staff on board. Their jokes may be corny, but they do try to be helpful.

VIGNETTE 11

Top Inventions—Only in Canada, Eh!

CANADIANS ARE VERY talented and innovative. The early settlers needed to be, just to survive, but many Canadians have made huge contributions to society. The following information dug from the Internet is a list of items invented by both famous and obscure Canadians:

1. Insulin, a life-saving treatment for diabetes, was the brainchild of Frederick Banting. Dr. Banting, along with colleagues, isolated the compound in 1921. At the time, diabetes was as deadly and disastrous as cancer. Banting was awarded the Nobel Prize in 1923.
2. Telephone. While the inventor of the telephone, Alexander Graham Bell, was born in Scotland, he immigrated to Canada as a young man and later set about to create a means to communicate across the long distances of his new country.
3. Light Bulb. Thought it was Edison's bright idea? Nope. Two Canadians, Henry Woodward and Matthew Evans, patented the light bulb in 1875. Unfortunately, the duo didn't have the funds to produce and sell the light bulbs, so they sold their idea to a man named Thomas Edison. That wasn't such a brilliant idea, fellas.
4. Five Pin Bowling. Five-pin bowling is a bowling variant played primarily in Canada and was created by Thomas Ryan of Toronto

in 1909. It was devised to offer bowlers the chance to play a quick game during a half-hour lunch break. This goal was achieved using smaller balls that can be cradled in the hand, travel faster than ten-pin balls, and can be thrown in rapid succession.

5. Wonder Bra. Louise Poirier of Canadelle Co. invented the push-up, bust enhancing Wonder Bra in 1964. When the Wonder Bra hit American shelves in 1994, it instantly became an American icon with obvious support.

6. The Pacemaker. While researching hypothermia, John Hopps discovered that you could restart a cooled heart with mechanical or electrical stimulation. He devised the first cardiac pacemaker in 1950, although initially it was much too large to be used internally.

7. Robertson screw. This special square headed screw and driver have a tighter fit than a slot screw, so that they rarely slip. Craftsmen soon found the Robertson screw to be superior to other screws, since it could be driven with one hand and was self-centering. But, while it is one of the most popular screws in its native Canada, it is almost unheard of outside of the Great White North because inventor Peter Robertson didn't pursue marketing his idea very aggressively.

8. The Zipper. Where would we be without the zipper? Cold and exposed? It was invented by Gideon Sundback in 1913, replacing cumbersome and unreliable fasteners like hooks and pins. And, Zippers are so much better at keeping out the rain, cold and snow than buttons.

9. Electric Wheelchair. After World War II, the influx of veterans that came home as para- or quadriplegics inspired George Klein, one of Canada's prolific inventors, to invent a motorized wheelchair.

10. Poutine. Ah, the Quebecers love their poutine. How can you not love a glorious mixture of French fries, melted cheese and cheese curds all smothered in hot gravy! An almost instant cardiac blood vessel blocker.

11. Basketball. James Naismith, the man with the extra empty peach baskets, wrote out the rules on only two sheets of paper and

Top Inventions—Only in Canada, Eh!

developed what was to become one of the most widely played sports in the world. And he wasn't even seven feet tall!

Honorable mentions include the CANADARM (outer-space giant reaching arm); Java Programming Language; Baby Pabulum, the Alkaline Battery; UV Degradable Plastics; Standard Time; Walkie Talkies; the Electric Prosthetic Hand; Kerosene; Electric Streetcars; Lacrosse; and Superman.

I read somewhere that Canadian inventors have patented more than a million inventions. In addition to those mentioned above, we can add Ginger Ale, Bombardier's "Ski-doo," Jacques Plante's hockey goalie mask, instant mashed potatoes, the lawn sprinkler, the plastic garbage bag and the electric oven. A more recent invention was the "Blackberry" developed by Mike Lazardis and Jim Balsillie and their team from Research in Motion.

VIGNETTE 12

History's Most Brilliant Inventions

I READ A short article recently that said one of the greatest achievements of the human mind is calculus. This person felt that calculus deserves a place in the honor role of our achievements with Shakespeare's plays, Beethoven's symphonies and Einstein's theory of relativity. Calculus, invented or developed separately by Newton and Leibniz, is said to be one of the best strategies ever developed for analyzing our world, and has made it possible to build bridges that span several miles of river, travel to the moon and predict patterns of population change, among other things. I wondered what else might be on a list of human inventions and checked the Internet to see what others thought. Personally, I lean towards lasagna, but some research yielded the following top ten list.

First place goes to the telephone, invented by Alexander Graham Bell in 1875. The telephone converts voice and sound signals into electrical impulses for transmission by wire to a different location, where another telephone receives the electrical impulses and turns them back into recognizable sounds. Number two is the computer, which has changed all of our lives. It is said that Konrad Zuse built the first freely programmable computer in 1936.

Television is said to be number three. In 1884, Paul Nipkow sent images over wires using a rotating metal disk technology with 18 lines of resolution. We've come a long way since then, given the microchip, HD, flat screens and many other developments. Perhaps the remote control should also have a place on this list.

Number four is the automobile. The very first self-propelled road vehicle (steam powered) was invented by French mechanic Nicolas Joseph Cugnot in 1769, but in 1885 Karl Benz designed and built the world's first practical automobile to be powered by an internal-combustion engine, and shortly thereafter Gottlieb Daimler took the internal combustion engine a step further and patented what is generally recognized as the prototype of the modern gas engine. Later he built the world's first four-wheeled motor vehicle. Henry Ford was the guy largely responsible for making the car accessible to us common folk, at least in North America. Some have actually described him as the most influential person of the 20th century.

The cotton gin, patented by Eli Whitney in 1794, was listed as number five. The cotton gin separates seeds, hulls and other unwanted materials from cotton after it has been picked. I've seen this item on many top invention lists and never understood why it deserves such honor, ahead of things like the printing press or the pacemaker.

Number six was the camera. The first camera was developed by Joseph Niépce in 1814, but the image required eight hours of light exposure, so instead a lad named Daguerre is considered the inventor of the first practical process of photography in 1837. Some guy named Kodak had some input as well.

The steam engine was number seven. Thomas Savery patented the first crude steam engine in 1698 and in 1765 James Watt improved Thomas Newcomen's design and invented what is considered the first modern steam engine. Number eight is the humble sewing machine. The first functional sewing machine was invented by the French tailor, Barthelemy Thimonnier, in 1830. Elias Howe patented the first lockstitch sewing machine in 1846 and Isaac Singer invented the up-and-down motion mechanism.

How about the light bulb at number nine! Humphry Davy, an English chemist, had some bright ideas as early as 1809. Contrary to popular belief, Thomas Edison didn't "invent" the light bulb, but rather he used some ideas from a couple of Canadians and improved upon a 50-year-old idea by developing a carbon filament that burned for forty hours.

Penicillin was listed as number ten. Alexander Fleming discovered penicillin in 1928, resulting in the ability to save millions of lives in war and in peace.

That's a pretty impressive list! I'm sure you could develop an equally impressive list that would be totally different. The ingenuity of humans is rather impressive as well.

VIGNETTE 13

The Red Lantern

IT'S CALLED THE Red Lantern, or in French *"La Lanterne Rouge"* or more colloquially, "the Tail Light". It's part of the "Tour de France", easily the most famous bicycle race in the world. This is a huge test of endurance where cyclists race for about three weeks over 2,000 miles (3,500 kilometers). The three weeks usually include at least two rest days, which are sometimes used to transport riders from a finish in one town to the start the next day in another town. The race attracts riders and teams from around the world; typically there are about 20 teams with nine members each. Since 1975 the finish has been on the Champs-Élysées in Paris.

The race is broken into day-long segments, called stages. Each day the winner of that stage wears the famous yellow jersey. Most stages are in mainland France, although since the 1960s it has become common to visit nearby countries for one or two stages. Individual times to finish each stage are aggregated to determine the overall winner at the end of the race, and of course on the last day the rider with the best cumulative time wears this coveted yellow jersey and receives a monetary award. Wearing it also confers some interesting power—for example if the race leader has to use the "roadside bathroom" during the race, the pack slows down to wait for him (although you never see this on TV!).

Most riders are part of the "general classification". While the yellow jersey is the most prestigious, there are other classifications or categories of awards. A white jersey is awarded to the best young rider. The green jersey is awarded to the rider who is best sprinter, and the best riders in this category often win a number of stages as they furiously attack each other near the end of the race in an effort to be the first across the finish line. They may not be as capable in cycling up the mountains, however, and the "King of the Mountain" polka dot jersey is given to those riders who consistently are the fastest mountain "climbers".

The mountain stages separate the pretenders from the professionals. The exact route varies a bit each year, but there always are challenging mountain climbs ranked from #4 (least difficult) to #1 (most difficult) based on length and steepness, and in addition there may be one or two mountain climbs that are labeled "*hors de categorie*" which means they are so difficult that they don't fit the normal classification. Points are awarded to those riders who are the first to reach each summit, based on the level of difficulty and the time taken to reach each summit. " *L'Alpe de Huez*" is probably the most famous mountain stage; it has an 8 percent grade for at least 14 km, with 21 hair pin turns, and depending on the route the stage that day can be 200 km long. The "*Col de Tourmalet*" has a 14 percent grade, another stage has an 8 percent grade for 15 km and several other climbs on a day when they ride 140 km, while another mountain stage has 40 km of uphill that includes five different climbs. So the "King of the Mountain" works very hard for the garish polka dot jersey.

In addition to the mountain stages (in the Pyrenees and the Alps) there are a number of stages that feature flat and rolling terrain; usually these are the longest stages. In recent years the Tour de France has started as a "Prologue", a time trial where riders leave the starting point one at a time every 30 seconds, and of course the fastest time wins that stage. There often are one or two more time trials dispersed during the race. The time trial is a different test from the regular stages since here they race individually against the clock compared to the regular stages, where most of the riders are bunched together as they race with their teammates. One reason they race so closely together in these stages is due to the power of the "peloton".

One thing that fascinates me about a bicycle race like the Tour de France is the power of the peloton. Peloton is a French word that means *little ball* or *platoon*. It is related to the English word pellet. The peloton is the pack or main group of riders in the race. Riders in a group save energy by riding close behind other riders (drafting or slipstreaming). The reduction in drag is dramatic; in the middle of a well-developed group it can be as much as 40 percent. Often a small group of riders will break away from the peloton to try to win the race, but if the peloton organizes itself by having different riders take turns in the lead, almost invariably it catches the breakaway riders, since the peloton is just too strong. Then, quite near the end of the stage, the sprinters in the peloton try to be the first across the line.

Professional cyclists are extremely fit, and the speed they maintain is incredible, even uphill. They average 40 to 50 kilometers per hour on flat stretches and usually over 30 kph up the hills and may burn around 5,000 calories in a four or five hour ride. A novice like me is fortunate to average 20–25 kph (12–15 miles per hour) and 7–12 kph up a steep hill while burning perhaps 300 calories per hour, and 60 to 80 kilometers is a long ride.

Oh yes, you were wondering about the Red Lantern. This is awarded to the cyclist who finishes last overall at the end of the race. You probably have guessed that only the best cyclists in the world are invited to be part of a Tour de France racing team. A key to the Red Lantern is that the awardee must *finish* the race (it is not unusual for up to 20 percent of the cyclists to "abandon" the race due to injury (crashes in the peloton are not uncommon) or other reasons (for example if a racer is too far back when a stage finishes they are disqualified from further stages). The winner of Lanterne Rouge has no hope of winning any of the prestigious jerseys, but they must have the courage and grit to finish the race. I like the metaphor of the taillight, which has the concept of being not too far from the headlight but is clearly still at the back. After this "award" gained some prominence, a few riders deliberately tried to finish last to be the taillight, so race organizers had to tighten the rules.

The pressure to win the Tour de France is very great. The reputation of the race has been severely tarnished for many years because of drug scandals. In the early days some said that it wasn't possible for the human body to ride 3500 km over 20 days on a tough course that included mountain climbs, and some riders used heroin, cocaine and strychnine to deal with the pain of riding huge distances. Amphetamines were commonly used as stimulants, and in 1967 a rider died from heart failure, apparently due to an over-dose of amphetamine. Race organizers basically turned a blind eye to drug use even after that tragedy, but finally, when the existence of the tour itself was threatened they took action and started serious testing. In 1998 one race team individual was found with a huge cache of growth hormone, testosterone and the blood enhancing drug EPO. At least two or three Tour de France winners have been implicated in drug usage. Lance Armstrong, a seven time winner, has always been subject to rumours of drug use, but despite being tested more than anyone, has never tested positive.

VIGNETTE 14

Restorative Justice, Forgiveness and Compassion

A SITUATION THAT I hope never to experience is one seen regularly on the TV news where an anguished and angry family comes out of a court house and is asked how they feel after a person is found guilty (or not) of killing one of their family members (and seemingly most often when the person is found guilty they are given what the family considers to be a light sentence). Often these folks want "justice" as they see it, and sometimes sadly and almost understandably, vengeance. "Justice" in these types of cases used to be capital punishment, but most Canadians would seem to settle for "life with no chance for parole". Rarely are the victim families able to demonstrate forgiveness.

People rightly feel that our justice system is skewed so that the "rights" of the victims aren't nearly as important as the rights of the offenders, or perhaps that judges and lawyers do just fine in the current system but no one else does. Law abiding citizens often feel that prison inmates "have it too good," and society generally believes that committing serious crimes must lead to the withdrawal of some basic rights and freedoms. But even in medium and minimum security institutions, if you look carefully, the loss of freedom is pretty dramatic. Still, prison guards and administrators have a huge responsibility to treat inmates as persons and to work effectively to rehabilitate them where this is possible. Do prisoners "deserve" pay (in 2011 about $6 a day) for the work they do while incarcerated? Do they use this money to buy a TV or books or shaving cream or hairspray? Do they save this money to help rebuild their lives when they are released?

I've had some experience with M2/W2, an organization which encourages society members to visit men and women in prison. In addition to providing friendship and outside contact for prisoners who often have been cut off from their families, one of the foundational basics of M2/W2 is the concept of restorative justice. That is, they don't advocate punishment or vengeance, but restoration of both the guilty and the victim, and society in general. One of the insights of restorative justice is that the need for healing goes beyond the victims (and the victims' families) and the offender, since the broader community is also damaged by criminal activity. Healing is not complete until the whole community is engaged and feels safe. Restorative justice conferences bring victims together with offenders and their families, and while they focus on the shamefulness of the offender's actions, they also affirm the inherent value of the offender as a person. Such interactions typically end with the offender agreeing to a specific course of action that will help him or her restore their relationship with the community.

A first step in restorative justice as a victim or family member is forgiving ourselves—this includes not feeling guilty about what we could have done or should not have done to prevent someone we love from being hurt, to the point where the guilt may inhibit our ability to function objectively, and feeling that we can love ourselves so that we can move on. The second step is to be able to love the perpetrator. We usually demand, however, that the perpetrator first exhibit remorse and then maybe, perhaps maybe, we can forgive them. But both victims and innocent bystanders can and need to take the first step in demonstrating care. A third step is letting go of all negative feelings and showing compassion. The guilty, even when they have done terrible things, and need to be "punished," are still human beings that need someone to care for them. The fourth step is the need to work toward life-restoring solutions. Restorative justice is constructive and healing. Hate is a terrible thing that hurts the "hater" as much or more than the "hated".

In restorative justice, dealing with those who break the law is still important. Telling the truth is the key—naming the wrongs, uncovering the truth and identifying root causes, and restoring the dignity of all parties. The fifth and last step in restorative justice is recovering the ability to trust and have compassion again by doing the right thing. We need to do something good with what happens to us rather than wallow in hate and self-pity. Even when we feel numbness and shock because of what has happened, we need to learn from the harm done to us. There is more to justice than punishment; there's restoration as well.

All of this is much easier said than done.

VIGNETTE 15

Should We Attach an Asterisk to Your Performance?

ACADEMICS ARE FAMILIAR with the concept of the asterisk or footnote to explain or document some particular observation or detail. In recent years this has also become an issue in baseball and other sports. Some first advocated placing an asterisk after Roger Maris' name since he took more than 154 games in a season to break Babe Ruth's iconic home run record. Then the issue of steroid-enhanced performance became a huge issue when stars like Mark McGuire, Sammy Sosa, Barry Bonds and Roger Clemens were thought to have been on some special juice to gain fame and fortune while dramatically setting records. There's a huge discussion going on regarding a number of athletes who have superb records that would apparently deserve membership in their sport's Hall of Fame, but who are suspected of cheating by using illegal drugs. It also seems clear that the East German Olympic Teams from 1976 to 1988 should have an asterisk attached to their records since it was subsequently established that many of their athletes, with their coaches' blessings and pressures, had used performance-enhancing drugs.

This is a pretty complicated issue, and making comparisons between players who played in different eras is fraught with complicating factors, even setting aside the ethical issues of using drugs to enhance performance. Some baseball experts would suggest that it's Babe Ruth's record that should have an asterisk since he never had to play at night or fly across the country every few days or bat against African

Should We Attach an Asterisk to Your Performance?

Americans who were kept out of the major leagues at that time. But then modern players have benefitted by Lasik eye surgery or modern training techniques or nutritional supplements or other modern day changes. Some athletes' records could have an asterisk to indicate that they could have done much better if they had taken better care of themselves and spent less time partying.

Perhaps we should consider using asterisks in ordinary life situations. How about putting an asterisk beside the name of anyone convicted of drunk driving, or those having more than three speeding tickets, or people caught texting while driving? Or perhaps beside the names of fathers delinquent in their child support payments? What about the students who have been found guilty of cheating in an exam? Perhaps asterisks beside the names of politicians who make promises but forget about them once they are elected? How about municipal politicians who commit expenditures way beyond the means of the municipality's capacity to balance the budget? Then there are those folks who routinely come to work late but make up for it by leaving early, but still expect a flattering performance review.

On occasion, those hosting "the Oscars" should have an asterisk behind their name to remind the organizers not to invite them again. For that matter, "The Academy" sometimes deserves an asterisk when a particularly odd choice is made. How about the hockey player that knocked Sidney Crosby out of action for at least one season plus playoffs, possibly ending his career? Or the mediocre minds that mete out punishment in the NHL in an unexplainable fashion? Perhaps one of the most infamous speeches by a politician in Canada is the 1995 speech by the slightly drunk Jacques Parizeau as he "conceded" losing the referendum on Quebec sovereignty; I think he merited two asterisks behind his name. I thought Kim Campbell would have an asterisk behind her name as Canada's shortest serving Prime Minister in 1993, but it turns out that this honor belongs to Charles Tupper way back in 1896.

One columnist wryly suggested setting up an "Asterisk Board" that would investigate situations like those just mentioned, and then put these names on a web site, plus provide opportunity for people to make recommendations for additions to the asterisk list. But it's always easy to see the deficiencies in others while ignoring our own foibles. I knew a fellow who was vehemently critical of government waste and inefficiency, yet he was cheating on his income tax returns. Like most of us, he believed that asterisks are for other people. Perhaps if all of us had more integrity we wouldn't need to think about asterisks.

VIGNETTE 16

Can You Say "Merry Christmas"?

SHOULD ATHEISTS, OR anyone that doesn't consider themselves to be a Christian, enjoy a celebration that stems from Christian beliefs? To celebrate or not celebrate Christmas is a decision that many serious non-believers are plagued with each year. Some say that it's hypocritical to be critical or pretty much disinterested in Christianity, and then acknowledge and celebrate a Christian holiday. But others suggest that Christmas isn't a holiday exclusively for Christians any more, and perhaps sadly, this has become true because our secular society has redefined Christmas. Some agnostics say that a Christmas tree simply symbolizes renewal, in contrast to the Christian symbolism supposedly introduced by Martin Luther who is said to have first decorated Christmas trees in honor of Christ's birth.

The purpose of Christmas for Christians is to celebrate the birth of Jesus. For others, it is a public holiday to celebrate with family and friends and to give and receive gifts. The earliest known reference to the date of the nativity as December 25 is found in a manuscript compiled in Rome in 354 AD. In the East, early Christians celebrated the birth of Christ as part of Epiphany (January 6), although this festival also emphasized celebration of the confirmation by the (three) wise men of Jesus as the one true God. The Christmas feast was introduced to Constantinople in 379. Around the 12th century, various traditions became accepted as part of the Twelve Days of Christmas (December 25–January 5).

In 1843, Charles Dickens wrote the novel *A Christmas Carol* that helped revive the 'spirit' of Christmas and seasonal celebrations. The instant popularity of this story played a major role in portraying Christmas as a holiday emphasizing family, goodwill, and compassion. In 1822, Clement Clarke Moore wrote the poem popularly known by its first line "*Twas the Night Before Christmas*", which helped popularize the tradition of exchanging gifts, and seasonal Christmas shopping soon began to assume economic importance. This also started the cultural conflict between the holiday's spiritual emphasis and its commercialism that some see today as corrupting the holiday.

Some young people don't want to tell their family that they have jettisoned their religious beliefs and some don't want to disrupt family traditions, so they'll do whatever makes their loved ones happy. In one school, given that there were many different types of religious backgrounds, out of respect they completely stopped celebrating ANY holiday, which seems like a huge overreaction. Others say that it's the message being repeated—the message of good wishes—that matters and that anyone who worries about being greeted with "Merry Christmas" is missing the point of carrying Christmas in their heart.

On one side are the politically correct who think that the expression "Merry Christmas" is too exclusionary, so they want us to say "Happy Holiday" or "Season's Greetings". They believe we should not say "Merry Christmas" because it might offend those of different faiths, like Muslims or Jews or Hindus, but, curiously it seems that it's politically correct ideologues and not these other religious groups that are offended. Many workplaces are insisting that employees use a generic greeting to promote intercultural respect and inclusiveness, and not to offend clients and other co-workers. With the well-intentioned interest of recognizing differences and wanting to be respectful of that difference, certain organizations have angered and touched a nerve for a lot of people by ignoring the origins and real purposes of Christmas. One person said that they say Happy Hanukkah to their Jewish friends on their special day, and they expect them to acknowledge and appreciate Christmas in return.

One secular viewpoint has been expressed this way: What is the holiday really about? The answer often given is, well, it is about shopping, about stores being open late, about getting the best sales so that you can buy gifts for people—not because you want to, but because that is what you have to do, because that's what you are made to feel like you have to do because everyone else is doing it. The season, they ironically say, is about people over-spending, over-eating and about selfishness. So, why not just call it what it is? Perhaps you could say, "Happy maxing out the credit card season!"

On the other side are the traditional Christians who perceive that the politically correct are attacking their religious beliefs and the customs their country was built on, but even more importantly secular folks are missing out on the real joy that Christmas can provide. To the traditional Christians, the attacks on Christmas are an attempt to de-Christianize society and the sacred traditions they hold dear. Many Christians strongly object to Christmas being dumbed down to "Happy Holidays".

Being respectful of other individuals and other cultures does not mean negating or ignoring our own culture or beliefs, whether our belief system is Christian or secular.

VIGNETTE 17

I Learned Everything I Need To Know in Kindergarten, From My Grandchildren and the Easter Bunny

YOU'VE PROBABLY SEEN this list in books or the Internet or on posters about learning in kindergarten—including share, play fair, don't hit people, put things back where you found them, don't take things that aren't yours, say you're sorry when you hurt somebody, wash your hands before you eat, flush, watch out for traffic when you go out in the world, hold hands and stick together, and be aware of wonder. And then remember the Dick-and-Jane books and the first word you learned: the biggest word of all—LOOK. Everything you need to know is in there somewhere—the Golden Rule and love and basic sanitation, plus ecology and politics and equality and sane living.

Well, as any grandparent can tell you, even after a fulfilling career, raising children, various travels and so on, everything they needed to learn they learned from their grandchildren. The following list is modified from the one compiled by Dina Santorelli.

First, smile for no reason at all—at the mailman, the clerk at the checkout, the door-to-door canvasser and so on. Most of the time they will smile back and the world is a better place.

Slow Down. This doesn't refer so much to automobile drivers as it does to those whose life is frenetically fast paced. Read a book; play a game with someone, slow the pace down.

Take joy in the little things. Smile when you see raisins in bread pudding or clouds floating in a bright blue sky or see two people hold hands.

I Learned Everything I Need To Know in Kindergarten,
From My Grandchildren and the Easter Bunny

Love is blind. Overlook small issues. Tell someone they are beautiful.

Clarity comes with time. Kids often inherently know what's right; hopefully your grandchildren have learned from your kids.

Don't take everything so seriously. Dine at a fancy restaurant with kids; don't worry about not wearing the right clothes, don't fret about whether your meal will be acceptable.

Change is good. Change the environment or the situation to diffuse disagreements.

People are listening even when you think they are not. Kids, or adults, may repeat something you said years ago.

There's always a solution. Compromise when you can, be creative, think outside the box.

The heart will go on. When a visit is over and the goodbyes are said, the heart still remembers.

Here's another list: All I need to know could also be learned from the Easter Bunny!

Don't put all your eggs in one basket. Everyone needs a friend who is all ears. There's no such thing as too much candy. All work and no play can make you a basket case. A cute tail attracts a lot of attention. Everyone is entitled to a bad hair day. Let happy thoughts multiply like rabbits. Some body parts should be floppy. Keep your paws off of other people's jelly beans. Good things come in small, sugar coated packages. The grass is always greener in someone else's basket. To show your true colors, you have to come out of the shell. The best things in life are still sweet and gooey.

VIGNETTE 18

Senior Cycling Adventures

I PASSED MY "best before date" some time ago, but since I do a lot of cycling, I thought that I might do OK if I entered the cycling events at the BC Seniors' Games in Richmond in 2009. Well, there's sad news and good news. There was a 16.6 km time trial, a 60 km race and a 2 km hill climb. There were various "Competitive" categories (60–65; 65–69, 70–80 and over 80 for both men and women, plus there was a "Novice" class in each category for first-time participants like me.

Well, for the 60 km race, which was 12 laps around a 5 km course, I finished dead last, *I mean like the very last rider.* The following were my initial observations about this event:

- Maybe I should have signed up for carpet bowling. Or horseshoes. Or bridge (yes, these were some other events).
- Wow, I guess it's true that some geezers are competitive and fit! I'm getting whiplash from the guys that are zooming past me.
- About lap six I remembered that, in the Tour de France there is a booby prize for last place, called "Le Lantern Rouge" roughly translates as "the taillight", but I didn't think they had such an award in this event.
- For the last few laps, it was getting lonely out here, on the circuit.

Some excellent and reasonable excuses for being last:

- My usual practice on a 50–70 km ride is to stop at a McDonalds at the half way point to refuel and read the paper before heading back home, but there was no McDonalds on this circuit.
- I had been away the previous week in Philadelphia eating steak and lasagna while teaching a laboratory Quality Assurance course.
- Even though I often go on a 60 km ride, I didn't train for a 60 km race.
- My bike isn't a racing bike.
- The other guys went so fast that drafting was impossible.
- I wasn't wearing spandex.

Some indications that I might be last:

- The RCMP traffic officer manning one of the corners said to me on lap 11 (since I was all by myself) "Sir, if you're not in this race you shouldn't be on this road".
- As I finished lap 11 the folks counting laps waved me down and told me to stop (but I did lap 12 anyway).
- The medal ceremonies (and it took them some time to check all the results first) were in progress as I finished the race.

Lessons learned:
- Don't play with the big dawgs unless you are a big dawg.
- If you do decide to play with the big dawgs, use the same quality and type of equipment. The fellow next to me said his bike cost only $2100, perhaps less than half of some of the bikes there (my hybrid bike cost $700).
- If you ignore rules #1 and #2, wear colorful spandex so you at least look good.

Looking on the bright side
- I did finish 60 km, and on the same day as the competitive guys.
- I did lap two women two or three times (the fact that they were both in the over-80 category did minimize this accomplishment).

- When I registered I received a free water bottle, a complementary T-shirt, and $50 for travel expenses!
- On the third day when I checked in for the hill climb, I found that I had won a bronze medal in the Novice category in the time trial on Day 1 (I didn't know they awarded medals immediately after the race so I had gone home).

Now for more good news!

I did the time trial of 16.6 km (10 miles) in 34 minutes, which is something like 29 km per hour, which is as fast as I've ever gone (did I mention that I won a bronze medal for this?) for that length of time. Then on the hill climb of two km on Saturday on Marine Drive in Vancouver my time was seven minutes and 34 seconds. I stuck around for the award ceremonies this time, and sure enough, I took the podium with another bronze medal in the Novice category.

Anyway, I do feel good that I participated and did my best.

I thought that the above would give some perspective on the Geezer Games. It was a good learning experience. There are a lot of old folks who take exercise in general, and cycling in particular, very seriously. One fellow told me he rode 90 km every other day. I don't really need a $3,000 bike and I can be happy biking 55–65 km (return) to the nearby towns of Chilliwack or Aldergrove for breakfast. One can be in the Novice category only once, so I would need to get serious and get a different bike to compete in the "competitive" category, and I'm not likely to do either of those things. But it was fun. And I don't think that carpet bowling is for me just yet.

VIGNETTE 19

Some Jokes and Tongue-in-cheek Comments

SURFING THE INTERNET yields all kinds of fascinating information. Here are some examples:

Seven things you will rarely hear from a husband:

(1) Here dear, why don't you have the remote tonight? (2) Honey, can I get you something while I'm up? (3) Sweetie, could you come to the mall with me to pick out a pair of shoes? (4) Dear, I watch the hockey game primarily for all of the neat commercials. (5) Honey why don't you relax tonight and I'll make dinner. (6) Sweetie, why don't we go visit your mother this weekend? (7) Could I have a salad with that instead of fries?

Reasons why it's great to be a guy:

You get credit for even the slightest demonstration of thoughtfulness.
Three pairs of shoes are plenty.
Your underwear only costs $10 for a three-pack.
You only need to know 5 colors.
Telephone calls can last less than a minute.
You can open all of your own jars.
You don't need to stop and think which way to turn the nut off of a bolt.

You can do your nails with a pocket knife.

When you go visit someone, you don't need to take a little present.

Wedding plans take care of themselves. (Plus, a wedding dress costs $5000, while tux rental, including shoes, might be $200).

You don't mooch off other people's desserts.

You can quietly watch a game with your buddy without ever thinking: "He must be mad at me."

If another guy shows up at the party in the same outfit, you could still become friends.

Your pals can be trusted never to trap you with "So, notice anything different?"

You are unable to see wrinkles in your clothes.

You might be an engineer if ...

Buying flowers for your wife or spending the money to upgrade your computer is a moral dilemma.

Everyone else on the Alaskan cruise is on deck peering at the scenery, and you are still on a personal tour of the engine room.

The salespeople at Circuit City can't answer any of your questions.

You are next in line on death row in a French prison when you find that the guillotine is not working properly, so you offer to fix it.

You can't fit any more colored pens in your shirt pocket.

In college you thought Spring Break was a metal fatigue failure.

On vacation, you are reading a computer manual and turning the pages faster than someone else who is reading a John Grisham novel.

You can't write unless the paper has both horizontal and vertical lines.

You find yourself at the airport on your vacation studying the baggage handling equipment.

You rotate your screen savers more frequently than your automobile tires.

You're in the back seat of your car with your wife, she's looking wistfully at the moon, and you're trying to locate a geosynchronous satellite.

Your three-year-old son asks why the sky is blue and you try to explain atmospheric absorption theory.

Some Jokes and Tongue-in-cheek Comments

Your wrist watch has more computing power than your neighbor's desktop computer.

Idle Thoughts of a Retired Person Whose Mind Wanders …

- I planted some bird seed. A bird came up. Now I don't know what to feed it.
- I had amnesia once—or twice.
- I used to be indecisive. Now I'm not sure.
- I went to San Francisco. I found someone's heart. Now what?
- Protons have mass? I didn't even know they were Catholic.
- All I ask is a chance to prove that money can't make me happy.
- What is a "free" gift? Aren't all gifts free?
- They told me I was gullible … And I believed them.
- Teach a child to be polite and courteous and when he grows up; he won't be able to merge his car onto a freeway.
- Two can live as cheaply as one, for half as long.
- How can there be self-help "groups"?
- One nice thing about egotists: they don't talk about other people.
- My weight is perfect for my height—which varies.
- Is it my imagination, or do Buffalo wings taste like chicken?

VIGNETTE 20

Airbus 380, the Boeing 747, and Other Heavier than Air Birds

I VISITED THE Boeing manufacturing plant near Everett, Washington a few years ago, and at that time they were trumpeting the introduction of the totally new Boeing 787 Dreamliner. There have been massive delays so I checked recently to see what Boeing was doing and what other developments there were in the aircraft industry.

The European built Airbus 380 is the largest passenger jet in the world. At about 79 feet tall it's as high as an eight-story building, and its wing span of 262 feet and length of 239 feet almost match the length of a football field. Its take-off weight is about 1.2 million pounds, which is considerably heavier than the Wright Brothers' "Kitty Hawk" in 1903 which weighed 625 pounds. Six international carriers, including Lufthansa, Air France, Singapore Airlines and Emirates, currently are using the A380 and have orders for many more. As of 2011 only five US airports are equipped to handle these massive planes, but this number will likely increase. Airports need extra loading bridges to load and unload passengers from the A380's upper and lower levels. Another airport challenge is the width of runways; there have been several incidents around the world where the wing tips of the A380 have clipped other planes or buildings, so that officials now require new runways to be 200 feet wide as compared to the standard 150 feet if airports wish to accept the A380.

Airbus 380, the Boeing 747, and Other Heavier than Air Birds

The A380 has several advantages for long haul flights. One A-380 carrying about 530 passengers can take the place of two average-sized airplanes carrying 250–300 people, thereby saving on fuel and pilot costs. The A380 is also said to be quieter and have better air quality than its competitors. Like other airplanes being manufactured today, the A380 uses carbon fiber composite materials for at least 25 percent of the plane's body, making it both lighter and stronger, as well as more fuel efficient and more environmentally friendly.

U.S. based Boeing is still in the picture, both for long haul international flights and shorter domestic flights. The first jumbo jet, the Boeing 747, is still widely used on long flights and is currently able to access more U.S. airports than the A380. Boeing has developed a new jumbo 747 that incorporates all the new technologies available to make it lighter and more fuel efficient. Boeing has also heavily invested in its new 787 Dreamliner which carries 210–290 passengers depending on the specific model, with a range of of about 8,000 miles (15,000 km). Various U.S. based carriers are committed to use the 787 as well as the workhorse 737, particularly since they are more cost effective for short hop flights within the U.S.

The Boeing 787 Dreamliner is described as a super-efficient airplane, and after a number of delays it is finally in production. A major advantage is that it brings big-jet range to mid-size airplanes, one reason being that the 787 has unmatched fuel efficiency, resulting in exceptional environmental performance. The airplane will use 20 percent less fuel for comparable missions than yesterday's similarly sized airplanes and will also travel at speeds similar to the jumbo jets. The interior environment with higher humidity will provide increased comfort and convenience. One of the new technologies is that almost 50 percent of the basic plane will be made of composite materials, and the design gives a simplified open architecture and increased functionality. General Electric and Rolls-Royce will provide the engines, which also will have increased efficiency.

While the A380 and the 747 (209 feet wingspan; 231 feet long) are huge passenger planes, they are not the largest airplanes in the world. The 6 engine Antonov An-225 Mriya cargo plane has a 288 feet wingspan and is 239 feet long; it's said to be the world's heaviest aircraft (but there is only one of them!). It was designed to airlift Russia's rocket boosters and the space shuttle for the Soviet space program. Howard Hughes' ill-fated Spruce Goose was 318 feet wide and 217 feet long, but it only flew once and is now just a curiosity.

By contrast, the world's smallest piloted airplane, the yellow and black "Bumble Bee", is only eight feet, 10 inches long and weighs 396 pounds, but it cruises at 150 mph. (Shorter, smaller versions were also built, but they crashed).

VIGNETTE 21

Mother Nature's Intricate Details

I ACCEPT THE idea, the fact, that God created the world and everything in it. I'm not a "young earth creationist" in that I believe that the world is 10 to 14 billion years old. But evolution just doesn't seem to be an adequate or defensible theory. Things do evolve, in that there is micro-evolution and there are minor variations, but in general the scientific evidence for evolution just doesn't stand up to scrutiny. Michael Behe, a biochemist, wrote an excellent book a few years ago called *Darwin's Black Box*, in which he says that when Darwin developed the theory of evolution the cell was a "black box" that no one understood. Now we understand pretty well the tremendous complexity of how the cell works, and that complex things only work if all the individual components work. If we take Behe's example of a mouse trap as a simple example, we need the base platform, a spring, a metal hammer which crushes the mouse, a sensitive catch that releases when slight pressure is applied, and a metal bar that holds the hammer back when the trap is charged. You need all five components for the trap to work; you can't start with just one or two or three components and let the trap "evolve" into a better trap. A physiological example is our body's complex blood clotting mechanism—every one of the dozen or so steps needs to work perfectly so that we don't bleed to death when we cut our finger; if only three or four steps work we're in trouble.

Big molecules that do the work in the cell, that is, proteins and nucleic acids, are composed of discrete units strung together. The building blocks of proteins are amino acids than can be fastened together to give an almost infinite number of proteins, but unless they connect in a specific (pre-ordained?) order the right proteins won't be synthesized and the cell won't work. DNA, the most important nucleic acid, is made up of four nucleotides that must be sequenced in specific ways, or else there will just be a mumbo jumbo and not a unique working building block.

God's accuracy and design may be observed in many aspects of nature, such as in the hatching of eggs. For example, the eggs of the potato bug hatch in seven days; those of the canary in 14 days; those of the barnyard hen in 21 days; the eggs of ducks and geese hatch in 28 days; those of the mallard in 35 days. The eggs of the parrot and the ostrich hatch in 42 days. (Notice, they are all divisible by seven, the number of days in a week!). Maybe not just a coincidence? God's wisdom is also seen in the making of an elephant—the four legs of this great beast all bend forward in the same direction. No other quadruped is so made. God planned that this animal would have a huge body too large to live on two legs, so for this reason He gave it four fulcrums so that it can rise from the ground easily. The horse rises from the ground on its two front legs first. A cow rises from the ground with its two hind legs first. How wise the Lord is in all His works of creation!

God's wisdom is revealed in His arrangement of sections and segments in many plants, as well as in the number of grains. For example, each watermelon has an even number of stripes on the rind; each orange has an even number of segments, each ear of corn has an even number of rows; each stalk of wheat has an even number of grains. Every bunch of bananas has on its lowest row an even number of bananas, and each row decreases by one, so that one row has an even number and the next row an odd number. All grains are found in even numbers on the stalks. All coincidence, or intelligent design?

The waves of the sea roll in on shore twenty-six times to the minute in all kinds of weather. God has caused the flowers to blossom at certain specified times during the day, so that Linnaeus, the great botanist, once said that if he had a conservatory containing the right kind of soil,

Mother Nature's Intricate Details

moisture and temperature, he could tell the time of day or night by the flowers that were open and those that were closed!

The lives of each person may be ordered by the Lord in a beautiful way for His glory, if you will only entrust Him with your life. If you try to regulate your own life, it will only be a mess and a failure. Only the One Who made the brain and the heart can successfully guide them to a profitable end.

VIGNETTE 22

The Human Body Is a Wonderful Thing

HAVE YOU EVER thought about how intricate the human body is? There are a zillion things that could go wrong as a foetus is developing in the womb and yet the vast majority of babies, all with a zillion connections and moving parts, are born healthy. Then, as a person grows, developmental and hormonal changes kick in at the right time as the baby becomes a toddler, then a young child, then a teen-ager with all of the requisite hormone additions, and finally an adult, and this individual unique person lives and breathes each day.

The human brain is obviously tremendously complex as well. I read recently about the exponential growth in the human ability to learn and store data. Computing power, which is the number of calculations per second available in all the computers in the world, is increasing at an average annual growth rate of over 50 percent. But here's the thing: Martin Hilbert at the University of Southern California, has stated that, despite the giant growth in digital calculating and storage capacity, the DNA in a single human body still stores far more information than all of the technology on earth! And the human brain computes far more calculations! He went on to say that the neurons in a single human brain fire more times per second than the number of calculations per second of all the computers on earth, indicating that computer techs are mere novices compared to Mother Nature. It seems that we are really "fearfully and wonderfully" made.

The Human Body Is a Wonderful Thing

Back to the intricacies of the human body. Think about what happens when you exercise. Your heart rate goes up fairly dramatically for a few minutes, but then, for some reason a control mechanism kicks in and it levels off. Good thing too, or we would literally blow our top. When you need to go to the bathroom, all it takes is a conscious thought and a valve opens so that you can relieve yourself. You know when you are thirsty because "osmo-receptors" sense a decrease in body fluid volume or an increase in electrolyte concentration, and messages are sent to the brain to let you know that you should drink some water or juice. This water balance happens with the complicated functioning of our kidneys, which are effective filters that work to conserve body fluid, or when necessary, ensure that excess fluid is excreted.

The intricate control of blood sugar levels is another example. Special cells in the pancreas produce insulin which, among other things, increases the rate of glucose absorption by tissue cells when blood levels are low. And how do you suppose it is that, when you stand up for a long time, the blood in your legs gets back up to your heart? Well, your heart pumps the blood into your arteries, some of which go down to your legs and into smaller and smaller vessels called capillaries where oxygen exchange into tissues (to allow them to function) takes place, and then the heart continues to pump the blood in veins back up to the heart. But there's a key feature of the vein that makes this flow against gravity possible—there are one-way valves in veins so that the blood keeps moving back toward the heart and the lungs to get a new supply of oxygen.

You may not have heard about the pineal gland, but this is a small endocrine gland in the brain that produces the serotonin derivative melatonin, a hormone that affects the modulation of waking and sleeping patterns and seasonal functions. We don't hear about it because it almost always works! The thyroid gland is also very complex and functions as part of a complex endocrine system, working under the regulation of the pituitary gland, which in turn functions under the regulation of the hypothalamus gland. The thyroid needs iodine to produce these hormones, and thyroid cells are the only cells in the human body that can absorb iodine. Sometimes, but relatively rarely, there are problems such as goiter (enlarged thyroid), hypothyroidism (underproduction of thyroid hormone), or hyperthyroidism (overproduction of thyroid hormone).

You have probably realized that there are many, many other examples. In all of these cases things sometimes can and do go wrong for some reason, but in the vast majority of cases our body works just like it was designed. If you watch the TV program "House" you may be fascinated by the "detective" work that Dr. House's team goes through to identify unusual medical problems. Dr. House is a bit unusual, to say the least, but the program illustrates how complicated the human body is, and how talented physicians can usually help make it function effectively in those cases where things go awry.

VIGNETTE 23

English Is a Crazy Language

I PARTICIPATED IN a Toastmasters Club for a few years, and this was very instructive in helping me avoid saying "uh," or "you know," or "well," or many of the other irritating fill-in words that people use when they don't quite know what to say. Participation in a Toastmasters Club is an excellent way to learn how to speak well, both in prepared speeches and extemporaneously. (One observation that I've made in watching programs like the "Oscars" is how poorly many of the actors speak in giving their acceptance speeches—without a script they stumble around just like many of us ordinary folks.)

Giving effective speeches is one thing, but learning English as a second language must be extraordinarily difficult. Permit me to provide a few examples from Toastmasters of our illogical use of words. A "near miss" is really a collision. Or, who's standing on a one night stand? If "extra-fine" means even finer than fine and "extra-large" means even larger than large, why doesn't "extraordinary" mean even more ordinary than ordinary? Why do they call food servers "waiters" when it's the customers who do the waiting? Why do we wear a "pair of pants," but not a "pair of shirts" (except on very cold days!)?

I heard once that French is an easy language to learn—three year olds in Quebec and France do it every day. Does that apply to English as well? If the plural of "tooth" is "teeth," shouldn't the plural of "booth" be "beeth"? If a school teacher has "taught," why hasn't a preacher "praught"? We say one "goose" and two "geese" but not one "moose" and two "meese". For some reason we can't get just one "jitter" or be in one "doldrum". A "slim chance" and a "fat chance" mean the same thing as do "good licking" and "bad licking"; and "caregiver" and "caretaker" are similar, but a "wise man" and a "wise guy" are very different. We often see signs that say "watch your head," and while this may be possible for some Star Wars characters, for us humans it's pretty much impossible. A fly can be a bothersome insect, or a verb describing what birds do or a metaphor for speed, or a noun defining something that men and boys should keep zipped.

Here are a few more questions: Why is it that when the sun or moon or stars are out they give light, but when the lights are out they are not visible? How can anyone understand me if "I'm speaking tongue in cheek"? If you say "put your best foot forward," how do you decide which foot goes first? And, if "fatty" means "lots of fat" why doesn't skinny mean "lots of skin"? Why do we often talk about things and ideas only when they are absent? For example, have you ever mentioned a "strapful" gown, described someone as "gruntled" or "combobulated" or "sheveled" or "ruly" or "peccable? I didn't think so.

Language evolves and is re-invented by people, not computers, and often English usage doesn't make sense. I could go on and on, but, well, uh, you know, that uh, we use many of these expressions every day. I won't venture into the abbreviations that "tweeters" and "texters" of the younger generation use since you would LOL at my pitiful attempts.

VIGNETTE 24

The Joy of Mathematics— An Oxymoron?

MATH SEEMS TO be a subject that one either hates or loves (I'm in the first category), and it often seems that we are either very good at it or very disinterested. Why is math important; why should a kid be required to take math in school? Well, because math is everywhere. Kids learn this early in life when they see their sibling or friend with more toys or cookies than they have, and when they get a loonie from Mom or Dad and need to figure out what they can buy.

Humans have been having fun with math for thousands of years. Math is important in every aspect of life—in science, engineering, finance, games of chance, in planning your monthly budget, keeping score in a ball game, or understanding the statistics you read in the newspaper. The next time that some polling pro says that candidate Jones has a 42 percent approval rating and that the results are valid 19 times out of 20, some math research will help you understand this. Math keeps planes in the air, makes credit card transactions secure, and powers your Google search engine. One professor describes math as "food for the brain" in that it helps one think precisely and creatively. If you are really keen on math, you may find a new way to experience beauty in the form of a surprising pattern or an elegant logical argument.

Perhaps a high school teacher's worst assignment is to be required to teach a remedial math class to disinterested students. Motivation is a key, especially for these kids, and an excellent teacher will strive to make things relevant to the kids. For example, studying Newton's equations of motion will be more interesting if the exercises are used to calculate the braking distance of a speeding BMW, or the distance that a human body would be thrown if not wearing a seatbelt. Mathematical models can also predict long-term impacts on the environment by analyzing available data. Math can also be used to help fight forest fires by predicting wind conditions and assessing the impact that changes in topography will have.

Some other real life examples where math is essential include search and rescue operations where math-based "tools" enable rescue teams to assess an area, and how best to get to where the injured person may be. Applied mathematicians are also making significant contributions in medicine. Math helps scientists map and understand the thousands of genes in our body to find the ones that may be responsible for a specific disease condition. In cancer research they can examine DNA and predict how cancer treatments will interact with patients' genes. Or they can examine MRI scans of brain cancer (which usually don't show the entire tumor) and use advanced geometry to show where the cancer has spread.

Understanding math is what keeps casinos in business—they have calculated the odds to make sure that the house almost always wins, and math makes tons of money for governments that regulate lotteries. Your friendly insurance company is another business where math is essential, since they know how much they need to collect in premiums to be able to pay out the claims that they calculate will be submitted. Mathematics is not just about making calculations, it's about thinking logically. Anyone paying interest on a mortgage or trying to understand a contract should make sure that they know how to do the necessary number crunching.

Calculus is a special type of counting; it's the branch of mathematics focused on limits, functions, derivatives, integrals and infinite series calculating change over time. In Latin, calculus means stone, since stones were used in ancient Rome to count and perform arithmetic. It is more advanced than algebra or geometry and is used to solve complex problems. Differential calculus is useful for calculating the speed of a rollercoaster, for example. Integral calculus, or integration, is used for more complex area and volume calculations such as the amount of water in a swimming pool. Credit card companies use differential calculus to determine the minimum payment required on a credit card. There are several variables that go into the calculation because it is determined by the amount of money that is due by the due date, plus the interest rate and the interest that has accrued since the last payment. With all the changing parts, interest rates and available balances, the calculations have to be done simultaneously in order to provide the customer with an accurate minimum payment.

VIGNETTE 25

Technologies that Changed the World

WHY DID THE U.S. go to the moon in 1969? Why did Edmund Hillary first climb Mt Everest? Why did Henry Ford produce an automobile that ordinary folks could afford? The answer for all of these questions is—because they could. Mankind has always had an insatiable desire to explore, to create things, and to make things better. I saw a list, compiled by Dr. Richard Lipsey, of "technologies that changed the world"; this is a list of ideas and things that made our world more productive, a list that reflects the incredible capabilities that we as humans have (we also have the ability to do terrible things to each other, but that's a different list and topic).

This list is chronological, starting with the domestication of plants about 9,000 BC and the domestication of animals. People no longer had to be nomadic hunter-gatherers, since they could choose to be farmers. The smelting of ore about 8,000 BC had a huge impact.

The invention of the wheel, about 4,000 BC, is often described as one of mankind's greatest technical achievements. Writing is said to have originated about 3,400 BC.

Bronze was first used about 2,800 BC while Iron was manufactured first around 1,200 BC. The Waterwheel was developed in the early Medieval Period, an important step in enhancing the ability and power of man.

Technologies that Changed the World

The 3-masted sailing ship first appeared in the 15th century, a significant step in facilitating global exploration and commerce. Printing, as in the invention of the printing press by Gutenberg about 1440, had a huge impact on the distribution of knowledge and in making education available to ordinary folks.

James Watt is usually credited with inventing the steam engine, but he actually just improved the invention by Thomas Savery who obtained a patent in 1698. Development of the factory system in the 18th century led directly to the industrial revolution and the mass production of products.

The "invention" of railroads is credited to Samuel Homfray in 1803 after he decided to fund the development of a steam-powered vehicle to replace the horse-drawn carts on the tramways or wagon ways in Great Britain. Progress in transportation reached a further huge step about the same time with the building of steam ships, and soon after that the internal combustion engine.

There seems to be some debate about who invented the internal combustion engine. Some say Nicolas-Joseph Cugnot invented the first self-propelled road vehicle in 1769. However the first gasoline engine was not created till 1859 by J. J. Étienne Lenoir, a French engineer, and Daimler took their concepts and developed the modern petrol (single cylinder) engine in 1885. He also developed the first gas operated motor-cycle. Carl Benz obtained the first patent for an automobile about the same time, but his first car appeared about 6 months later. Think about that when you are stuck in a traffic jam.

Electricity was not "invented" by any one person, but its properties were discovered through the research of many great thinkers like Benjamin Franklin, Thomas Edison, and Nikola Tesla. But Michael Faraday did invent the first electric motor. Next on our list is the airplane. Orville Wright flew in the first successful airplane flight at Kitty Hawk, N.C. on December 17, 1903, in a plane designed by his brother Wilbur and himself.

You knew the computer would be on the list. Charles Babbage invented the concept of a programmable computer-like device in about 1856. The ENIAC 1 computer, with 20,000 vacuum tubes, appeared in 1946. Transistors first appeared in 1947, thereby greatly facilitating the development of computers, as did the integrated circuit or chip in 1958. Microprocessors, floppy discs, flash drives—well the list goes on and on, and then came networking and the Internet.

Biotechnology and Nanotechnology complete our little tour of technology development. We will hear a lot more about these technologies in the coming days.

VIGNETTE 26

Everyday English and Interesting and Conflicting Expressions

COMMON EVERYDAY ENGLISH has some interesting expressions and some irritating phrases, especially if they are repeated over and over again. Why do people keep saying "at this point in time," or, "get on the same page," or, "think outside the box," or, "win-win situation"? The word "iconic" also seems to be over-used as an adjective in attempting to express the importance of a person or event.

And people love to mischievously or seriously say ridiculous things, like, "I wouldn't be paranoid if everybody didn't pick on me," or, "I'd give my right arm to be ambidextrous."

Some words are always nouns or always verbs, but some words do double duty. A monkey can be an animal or a verb meaning fooling around. A book can be something we read or that act of making a reservation, or placing a bet, or making an arrest. A table is a piece of furniture, but it also can mean the act of putting an item on a schedule. We water plants (a verb) with water (a noun). A monitor can be a screen attached to our computer, or the act of watching things.

Many English words have at least two very different meanings. A sound can be something we hear, or a body of water, or assessing what a person thinks. A post can be a stick in the ground, something we mail, or a diplomatic location. Tack can be something we use to pin paper to a board, or harness and equipment for a horse. Cookies are calorie rich treats or pieces of computer code. How about shingles—a nasty viral disease, or roofing material, or a way to cut a lady's hair? A ball may be a round thing to throw or hit, but also a fancy dancing event. Have you ever been at a bored meeting? Reefer is another interesting word; this is the slang word for a marijuana cigarette, but reefer can also refer to refrigerated trucks that haul food, for example, and less commonly reefer can also mean a thick double breasted jacket. Another unusual word is "habilitation" (or habilitate). I encountered this word when I saw a sign that said "Habilitation Center". We are familiar with rehabilitation, but habilitation can mean "to clothe or dress", or in the context just mentioned "to become fit or capable".

Some scholars estimate that there are over a million words in the English language, so why do we have to have two or three meanings for some words? Couldn't somebody invent a few more words? The estimates for the number of words that an average person uses is said to be about 5,000 although anyone with an expanded vocabulary may use about 10,000 to over 60,000 depending on education and background, while understanding as many as 250,000 words. Winston Churchill is said to have understood about 400,000 words.

And don't get me started on spelling and pronunciation. Why is solder pronounced as sodder? Why don't we spell a sharp weapon with a blade as "sord" instead of "sword"? Did the "K" have a problem when the alphabet was developed? Why don't we spell cute as kute or compost as kompost? Do we really need this letter? Sometimes it seems to start a word for no apparent reason, as in knock or knit or knee or even knight; there are many many more examples.

It's often very rewarding to read talented authors who use rich expressions and have flair with language usage. Some of my favorite examples include "the flames eagerly looked around for fuel," and "he climbed the steps with vexatious phlegm," and "morale was lower than a gopher's basement." And I like "she turned to me as a miser might look if anyone attempted to assist him in counting his gold," and "when a woman wants to say nothing, the silence can be deafening." I also like the quote by Joe Adcock about a famous home run hitter in baseball: "Trying to sneak a fastball past Hank Aaron is like trying to sneak a sunrise past a rooster."

VIGNETTE 27

Things I Remember as a Kid that Are Now Obsolete

PEOPLE, ESPECIALLY YOUNG people and seniors, love to make fun of old people, like not remembering what they came to do when they entered a room or making jokes like "everything hurts and what doesn't hurt doesn't work." But I'd like to take a more whimsical view of things by recalling items that were regularly used when I was growing up (which happened on a farm in Alberta). You may need someone over 65 to explain some of the things described here, depending on your vintage and background.

How about black and white TVs with only three channels and lots of "snow"? Or battery operated radios, with tubes that needed periodic replacement? VCR's had a brief and spectacularly short history. We still have a few 35 mm slides, but tossed out our slide projector after it stopped working. Did you ever see lantern slides at some meeting? Do you still use an Instamatic camera equipped with four flashbulbs? I was in a shop recently that had thousands of LP records. Records!!?? Do you remember "record players" or 45 rpm records? City folks make fun of outhouses, which were a common feature when houses didn't have indoor plumbing. In this regard, it seems cruel now, but Sears and Eaton's catalogues had a useful sanitary function besides serving as wish books. City folks had a milkman deliver milk in glass bottles.

My mother used a flat iron with a detachable wood handle, so that the second iron could be warmed on the stove while the first one was being used. And yes, it was a black coal and wood burning stove. Mom also used a gas powered ringer washing machine before electricity finally came to central Alberta in the mid-1950s. A washboard was used for scrubbing particularly badly soiled garments. Many modern ladies may not even know about sewing machines, and especially not about foot-treadle sewing machines. No more hanging on to mother's apron strings because she doesn't wear aprons anymore!

Automobiles used to come with a crank for starting when the battery wasn't working. I can remember when gas was the extravagant price of 39 cents per gallon. The attendant (you remember gas jockeys?) used to hand pump the gas from the underground storage tank to a big bulb or glass tank above the gas "pump," and then gravity took over as the gas flowed through the hose into the car's gas tank. This was when engines had carburetors. Flat tires are now pretty much obsolete, except when caused by vandals. In fact, some manufacturers are now selling cars without a spare tire or a jack or a tire wrench, although they are kind enough to give you the privilege of paying extra for a "compact spare" as an option. Studebakers looked weird even back then.

Typewriters? My granddaughter found our old electric typewriter recently and wondered what it was; she should have seen the original manual typewriters with the carriage return and carbons! Speaking of carbons, have you ever seen a Gestetner machine? Do you remember chalkboards at school and the dusty erasers? We even had a giant pot-bellied stove at the back of the school classroom. When I was in Grade 4 we learned to write with refillable fountain pens; I think every kid in the class spilled their bottle of ink at least once during the year. As we got older we discovered drive-in movie theatres, among other things. And on Sundays we heard the church bells pealing away.

Well, you get the picture. The times, and things, they are "a changing." Sometimes that's very good, like doing away with the smoking section in airplanes. Will we add the gas powered internal combustion engine to the obsolete list someday? Or books?! Perhaps single family homes, except for the super wealthy, will soon be history, since the sales of condominiums are increasing much more rapidly, while fewer folks are able to buy houses.

VIGNETTE 28

Are Computers Male or Female?

ONE OF THE first unfathomable things that I learned in high school French was that French nouns, unlike their English counterparts, are grammatically designated as masculine or feminine. A house is feminine (la maison), a book is masculine (le livre), table is feminine, and so on. I suppose that linguistics experts know why this is, but it's one of life's little mysteries that I don't feel a need to understand. Assigning gender to nouns probably happens in other languages as well, but then I know even less about Swahili or other languages where this might be the rule. Some wise student today might wonder what gender is a computer? The internet, as always, has the answer.

A group of female students was asked this question, and they concluded that computers should be of the masculine gender. They offered these arguments to support their conclusion: in order to get their attention, you have to turn them on; they have a lot of data but are still pretty much clueless; they are supposed to help you with your problems, but half the time they are the problem; and, as soon as you commit to one you realize that if you had waited a little longer you might have had a better one. On the other hand, the male students said that computers were clearly of the feminine gender because no one but their creator understands their internal logic; even your smallest mistakes are stored in long term memory; the native language they use to communicate with other computers is incomprehensible to everyone else; and, as soon as you make a commitment to one you find yourself spending half your pay check buying accessories for it. Given the characteristics of both women and computers, the statement "men are from Mars and women are from Venus" clearly applies here.

The idiosyncrasies or variable performance of computers, particularly in their early history, resulted in an interesting exchange some years ago when Bill Gates reportedly criticized the auto industry for not keeping up technologically as quickly as did the computer industry. The response from General Motors went something like this: If GM had developed technology like Microsoft did, we would all be driving cars with the following characteristics:

For no reason whatsoever your car would crash periodically or even frequently. Every time they repainted the lines on the road you would have to buy a new car. Occasionally your car would die on the freeway for no reason, and you would just accept this, restart and drive on. When executing a manoeuvre such as a left turn, your car would sometimes shut down and refuse to restart, in which case you would have to reinstall the engine. Macintosh would make a car that was powered by the sun, was reliable, faster, and easier to drive, but would only run on 5 percent of the roads. The oil, water temperature and alternator warning lights would be replaced by a single "general car default" warning light. The airbag system would say, "Are you sure?" before going off. Occasionally for no reason whatsoever, your car would lock you out and refuse to let you in until you simultaneously lifted the door handle, turned the key, and grabbed hold of the radio antenna. Every time GM introduced a new model car buyers would have to learn how to drive all over again, because none of the controls would operate in the same manner as the old car.

We love computers, and ladies, and life would be much less interesting and productive without either of them!

VIGNETTE 29

Some Comic Strips Are Hilarious; Others Not So Much

WHEN I READ a newspaper I only turn to the news and sports sections after reading the comic section. Perhaps I'm trying to avoid reality, but my thought is that I enjoy a bit a humor, and I'm sticking to this story. Comics often provide humorous, sometimes ironic, sometimes thought-provoking insights into how people approach the challenges and joys of everyday life.

My favorite comic strip of all time has to be "Blondie". Of course Dagwood is the real star, with two major talents (designing and eating massive delectable sandwiches, and procrastination, usually exhibited by napping on the couch). Perhaps his other main talents are frustrating his boss by sleeping or lolly gagging at work, and frustrating his carpool buddies by routinely being late. We can all relate to Dagwood's talents and his family discussions, which probably accounts for the enduring and endearing nature of this comic strip.

Another of my favorite comic strips is "Calvin and Hobbes", written by Bill Watterson, which describes the antics of Calvin, a precocious six-year-old, and Hobbes, his sardonic and anthropomorphic stuffed tiger. The names come from John Calvin, a 16th-century French Reformation theologian, and Thomas Hobbes, a 17th-century English political philosopher, and one often feels the tension between these philosophical perspectives. The strip describes Calvin's flights of fantasy and his friendship with Hobbes, as well Calvin's turbulent relationships with his parents and classmates. For Calvin, Hobbes is a live tiger with human thoughts, but everyone else sees him as an inanimate stuffed toy. Some of the strips show quite a dark side, but most are fun-loving descriptions of an adventurous boy, while others describe various levels of philosophy designed to make you both think and laugh. "Doonesbury", by Gary Trudeau, is another comic strip that makes you think, but the messages are very different. It critiques American war, culture and economic policies with a touch of sarcasm, satire and disbelief.

Of course I have to mention Charles Shultz's "Peanuts" and the two main stars Charlie Brown and Snoopy, even though this strip is no longer being produced. Our long suffering friend Charlie Brown exemplifies the struggles that many of us, especially as kids, go through. The Peanuts TV specials that block out all adult voices and just tell the kids' stories are a neat way of emphasizing the world of children and how they deal with life's tests and circumstances. The anthropomorphic stories of Snoopy, with his friends, and especially his dog house with the underground pool table and who knows what else, plus his eccentric brother Spike, present a wonderful sense of imagination and humor.

I think one of the best relatively new comic strips is "Betty", which effectively and humorously describes a small family dealing with modern day-to-day issues. Another family type strip is "Zits", which chronicles the foibles and challenges of a modern day teen-ager as he copes with his hopelessly outdated parents, as well as with his friends and something foreign called "responsibility". I also enjoy the family oriented comics like "Dennis the Menace", "Family Circus" and "Adam". "Heart of the City" is lots of fun as it humorously describes the vivid imagination and antics of a hyperactive little girl.

Comics that I don't particularly enjoy are "Speedbump", "Flying McCoys", "R Minus", "Animal Crackers" and "Monty", although they occasionally have witty and incisive issues and repartee. "Rudy Park", "Tina's Groove", "The Other Coast" and "Luann" are generally worth ignoring. "For Better or Worse", "Fred Basset", and "Sally Forth" seem to have lost their reason for existence and aren't very interesting to me anymore. I tried to read "Rex Morgan" for a couple of months, but finally gave up since it just seemed to be a non-essential soap opera. Comics that I no longer read at all include "Pearls Before Swine", "Cul de Sac" and

Some Comic Strips Are Hilarious; Others Not So Much

"Get Fuzzy", since they seem to be just too dumb and not very funny. The concept behind Dilbert, which seems to be to mock executives, may have been a good idea, but the everyday strips are just too silly to even be funny.

One of the best ever comic strips was "Pogo", since it had superb home spun philosophy and humor ("we have found the enemy, and they is us"); it's too bad that it is no longer being written. If you're really old, or eccentric, you may remember "Andy Capp", a lovable English lay-about with a fondness for brew and racing and an ever-so-patient wife. Another very old and obsolete comic strip was "Little Orphan Annie", the little girl who apparently had no eyes but saw many different poor kid adventures.

So, read the comics to enjoy some humor and some fun, and both a serious and a humorous view of life. Then, if you must, go read the news of the world.

VIGNETTE 30

Watch This!

WHEN I WAS a kid, lo these many years ago, it was the custom for some reason to give kids their first watch in Grade 9. These days, kids as young as five or six are sporting watches that can give you the time in all the major cities of the world, plus the temperature and your hair color. I checked the Internet to learn more about the history of watches. I have a friend who has more than a dozen watches. They say that a person with one watch always knows the time, but a person with two watches is never quite sure.

The wrist watch was invented by a Swiss watch maker in the late 1800s, though at first only women wore them. A wristwatch for men was invented in the early 1900s for an industrialist who found a pocket watch inconvenient to look at while doing his work. When we think of fine watches today the Swiss still come to mind, even though both the English and the Americans were once considered leaders in the watch field. The Swiss had a flair for the fashions that were desired by the rest of Europe and the world, and they made a wide range of watches, from cheap junk to very high quality complicated watches. Watches were made in small batches by small companies "by hand". Starting around 1850, the Americans pioneered the use of automated machines to mass produce high quality watches with interchangeable parts, and they soon proved that this system could make watches that were every bit as good as all but the very best watches that were made by hand, plus they were cheaper. Swiss and English watches virtually disappeared from the American market for a while. One drawback to the American system of manufacturing was that each part required a machine to make it, so complicated watches, chronographs and very high end watches were not practical to make.

The English basically folded up shop for a time and stuck to making high end watches and chronometers needed by British ships. The Swiss, on the other hand, reacted by adapting to the new market, and they reorganized into centralized factories, with a fair amount of automation and improvements so that they caught up with the Americans. They soon had many companies making various components like watch dials or mainsprings, which would be sold to other companies that would complete the watches and sell them. With so many Swiss companies each doing only a part of the manufacture, the Swiss were able to produce everything from very cheap watches to watches of the highest quality. This gave the Swiss more production flexibility when supply or manufacturing problems occurred. During WWII American watch companies sunk much of their capital into converting to war production, such as bomb fuses or timers and navigation equipment, and since the Swiss were "neutral" they were allowed back into the American market, which they dominated when the war ended. The American public now had money, and the Swiss were willing to supply them with watches. By the 1950s, the Swiss had perfected machine-made complicated wrist watches such as chronographs, automatic winding watches, and day-date watches. In 1926, the first self-winding wristwatch was produced, and the first electrical watches were introduced later in 1952.

The only American watch company that survived all of these changes was Timex, which made cheap, "disposable" watches. While they were looked down on by the elite watch makers, they were at least making a profit. By making the watches disposable, Timex was able to do things like completely seal the watch case, which meant that it couldn't be opened to be repaired, but it also wouldn't let dust in. The Timex watch also lacked any jewels, which meant that it would wear out after a while, but it was still rugged. A sharp knock would often break the jewels in an expensive watch, but a Timex could "Take a Licking, and Keep on Ticking!" The other American company of note was Bulova, which up until the 1950s had imported their movements from Switzerland. In the 1960s, they created the revolutionary electronic "Accutron" watch, which used a tuning fork to keep time instead of a rotating balance wheel, and the result was an incredibly accurate watch. The Accutron became the high end watch from the early 1960s until the early 1970s when the quartz watch took over.

In the 1970s, the Swiss had another shock when the Japanese perfected the quartz watch. Major technical developments soon followed, such as LED and LCD displays and the quartz wristwatch without battery. In 1982, the somewhat smug and superior Swiss horological establishment was in crisis, and their rich watchmaking tradition and splendid history of innovation was in danger of being swept away by some quirky material that vibrated at a particular frequency when captured within an electric field, namely quartz technology. The Japanese quartz invasion, and, to a lesser extent, the emergence of the American jewel-free, throw-away watch company Timex, was a dramatic blow to the Swiss, and many solid and cherished brands went out of business. But the Swiss are resilient and they adapted by hiring a consultant who knew little about the mass production of timepieces but who knew how to fix problems. Nicholas Heyek developed a turn-around plan for a major Swiss watch company and he led the Swiss out of their doldrums. The Swiss industry still wanted to make high quality watches, but Heyek's team established basic principles to also be able to sell watches for the lower end of the market; namely they must have style, they must be cheap to make, they must be priced competitively, and they must be durable. Their vision lead ultimately to the creation of the Swatch in 1983, a brilliant fusion of style and technology and a brand that gained back much of the ground lost to the Japanese. The number of parts used to produce the watch was reduced to around 60 percent of those in similar models, and economies were achieved by robotics and single assembly lines.

Do you have the time?

VIGNETTE 31

What Time Will It Be When We Get to Toronto?

IN THE OLDEN days, as they used to say, time was determined by where the sun was in the sky. People used to set their clocks at high noon when the sun, more or less precisely, was directly overhead. Time was perceived and set differently at each different location and town. Time of day was a local matter, and most cities and towns used some form of local solar time, maintained by some well-known clock (for example, on a church steeple or in a jeweller's window). Travellers in 1870 had to change their time in each different city. When it was noon in Washington, DC it was 12.54 pm in Halifax, 12.14 pm in Montreal and 11.54 am in Toronto.

For a long time this worked pretty well since folks didn't travel very far or very fast. But when railroads came into existence in the 1800s, determining the correct time became both a safety issue as well as convenience issue. Moreover, it was very difficult to devise and follow schedules. This was a bit of a problem in Europe and it became a huge problem in North America where the distances involved were much greater.

A brilliant Canadian by the name of Sir Sandford Fleming proposed the concept of 24 standardized world time zones, still in use today. Sandford Fleming immigrated to Canada in 1845 and first worked as a surveyor and later became a railway engineer for the Canadian Pacific Railway. Standard time in time zones was instituted in the U.S. and Canada by the railroads in November 1883, and Fleming was also instrumental in convening the 1884 International Prime Meridian

Conference in Washington, which firmly established internationally the concept of time zones. Greenwich, England is known for its maritime history but especially for giving its name to the Greenwich Meridian (0° longitude) and Greenwich Mean Time. Fleming made another important contribution. In 1876 he missed a train he was supposed to take. The train's schedule had misprinted the departure time as p.m. instead of a.m., so Fleming was then inspired to create a 24-hour international clock.

Daylight saving time is the practice of temporarily advancing clocks during the summertime so that afternoons have more daylight and mornings have less. Spring forward, as they say, one hour in spring and fall back in autumn. Daylight Savings Time was first proposed in 1895 by George Vernon Hudson in New Zealand. Many countries have used it since then, but the practice has been both praised and criticized. Adding daylight to afternoons is said to benefit retailing, some sports, and other activities that exploit sunlight after working hours, but causes problems for farming, evening entertainment and other occupations tied to the sun. An early goal was to reduce evening usage of electricity, but the impact on energy use is limited or contradictory.

Daylight Savings Time presents other challenges. Who hasn't been an hour late or an hour early because of the time change, and meetings across several time zones can be even more complicated than usual when some areas have Daylight Savings Time and some don't. Travel, billing, recordkeeping, medical devices, heavy equipment, and sleep patterns can be disrupted. Modern software usually adjusts computer clocks automatically, but this can also be problematic.

It turns out that there is another complication in keeping time. According to my grandson's "Canadian Geosystems" textbook, the basis of marking time is that the earth revolves 360 degrees every 24 hours, but keeping accurate time is not just a matter of tracking the earth's rotation. It seems that the rotation of the earth is very gradually slowing down due to the tug of tidal forces. I should emphasize "gradually" since they say that in about 150 million years the length of a day could be as much as 27 hours. This projection will be difficult for you and me to verify, but you may have noticed that every so often "they" add a few seconds or portions thereof to our calendar year, so after 150 million years or so this could add up.

VIGNETTE 32

Keeping Time

A CLOCK IS defined as a device having two qualities; a regular, constant or repetitive process or action to mark off equal increments of time, and displaying the result. Today, more than ever, we are interested in keeping time. In some sports they measure time to the thousandths of a second; most employers are satisfied if their staff come to work on time or only minutes late. But keeping time goes back to the ancients. Celestial bodies like the sun, moon, planets, and stars have provided us a reference for measuring the passage of time throughout our existence. Ancient civilizations relied upon the apparent motion of these bodies through the sky to determine seasons, months, and years.

 The Greeks developed something called a clepsydra which worked by measuring the dripping of water. The Egyptians had candle clocks which worked well if the wax was consistent and the painted stripes were evenly marked. In the Orient, small stone or metal mazes filled with incense that would burn at a certain pace were invented. Then there's the sundial, which divides the day (which is longer in summer than in winter) so you can tell when midday is, but it doesn't work too well in Whitehorse or Fort St. John in winter! Around 3500 B.C., the Egyptians built obelisks—tall four-sided tapered monuments—and placed them in strategic locations to cast shadows from the sun. Their moving shadows formed a kind of sundial, enabling citizens to partition the day into two parts divided by noon. They also showed the year's longest and shortest days when the shadow at noon was the shortest or longest of the year. Later, markers added around the base of the monument would indicate further time subdivisions.

A pressing need for more precise measurements led to the invention of the water clock and then the hourglass, but these inventions were limited, as they relied on water or sand to function. The search continued for a way of tracking time independent of the seasons or nature.

In 1582, Galileo discovered that a pendulum could be used to track time. He drew the first designs for a clock, though he did not build it. Finally, in 1656, Christiaan Huygens built the first known clock, putting Galileo's discovery to use. Though the clock did not keep accurate time, it was a major breakthrough in timekeeping technology. Through the years, various inventors tried to improve on the design, but with little luck. Finally, in 1670, William Clement discovered that the clocks worked better with a longer pendulum. Of course, this required a taller clock. Clement named his newly designed clocks long-case clocks, which were the predecessors of the "grandfather clocks" that we know today. In 1875 Henry Work wrote the song "My Grandfather's Clock" and the name stuck and finally, after many years in the making, grandfather clocks were born.

Today's grandfather clocks rely on a pendulum attached to an anchor. The swinging pendulum causes the anchor to turn a gear, which in turn causes the clock to tick. A pair of weights further helps power the clock and keeps it from losing time. Although the technology has evolved over the years, the grandfather clocks of today still reflect the ingenuity of humans over the centuries.

Today, we think of pocket watches as quaint antiques or family heirlooms, but until World War I, they were the only portable source of timekeeping available. According to the internet, a German locksmith named Peter Henlein invented the first "watch" in the early 1500s. Until then, clocks were powered by falling weights and pendulums, which made it necessary for them to be stationary and upright to operate. Henlein created the mainspring, which enabled clocks to be portable so we could carry them around. About a hundred years later, Robert Hooke designed a watch with a balance spring, which controlled the oscillations of the wheel that more efficiently controlled the watch's operation. Using the mainspring, clockmakers could produce portable, small clocks that could be carried or worn on a chain. Most of the world's early watches came from Germany or France.

Early watches were made of iron and were quite heavy so they were suspended from a chain or cord and worn around the neck or hung from a belt. They were also usually inaccurate and only had an hour hand, until the minute hand was developed in the late 17th century. The second hand did not come along until the 20th century. Early watches were enclosed in cases shaped like spheres or cylinders, but more unusual-shaped cases, like skulls or crosses, became stylish in the mid-1600s.

Eventually watchmakers began using brass instead of iron, and watches became smaller and light enough to fit into the pocket of a jacket or vest. Pocket watches were handmade and fairly expensive until the 19th century, so they were considered luxury items and mostly worn by the upper class and merchants. Aaron Dennison was the first to adapt the concept of interchangeable parts to the production of pocket watches so they could be mass-produced. He formed the Waltham Watch Co company in 1849 and this was one of the most popular pocket watch makers until the 1950s.

VIGNETTE 33

My List of "Most Admired People"

THIS IS MY own list of (primarily) modern day people who have made significant contributions to society.

Mother Teresa usually tops lists like this, and deservedly so. She devoted her life to helping the poor in India.

Mom. I sincerely hope that you can put your Mom at or near the top of this list, as I can. Moms are pretty special and do so much for their kids.

Billy Graham is often near the top of admired people lists because of his integrity and his work to encourage people to become Christians.

Nelson Mandela deserves a high place on my list. After 27 years of imprisonment and government invective he could have been bitter and unleashed a torrent of unrest in South Africa when he was released, but he chose a better way to improve the life of his people. Archbishop Desmond Tutu also belongs here, in part because of his key and effective moderating role in the "Truth and Reconciliation" Commission.

Martin Luther King Jr. is rightly revered by most Americans and even has a national holiday named after him because of his non-violent and effective role in advocating American civil rights. We could also mention Rosa Parks who one day had the courage not to move to the back of the bus, and ordinary black Americans therefor owe her a debt of gratitude.

My List of "Most Admired People"

Albert Einstein's name is recognized around the world as one of the greatest physicists ever, since he developed the theory of relativity and a little equation deciding that "E equals m c^2".

Not all Popes have been loved or greatly respected, but Pope John Paul II certainly was after rising from humble Polish beginnings. Mohandas Gandhi is still revered because of his non-violent but hugely effective approach in working for Indian independence from British rule.

Tommy Douglas was voted a few years ago as the greatest Canadian by a CBC TV series on the basis of being the driving force for universal health care and his concern for his fellow citizens and especially the poor. Terry Fox, Wayne Gretzky, Alexander Graham Bell, and Frederick Banting (who discovered insulin) made that CBC list of great Canadians, and mine.

John A. Macdonald and Wilfrid Laurier were respected Canadian Prime Ministers, while Lester Pearson was average as a Prime Minister but hugely admired for his external affairs expertise and was rewarded with a Nobel Peace Prize for suggesting the idea of UN Peace Keepers.

I always admire brilliant people who use their gifts and talents in a positive way. This applies to C.S Lewis, author of the *Narnia Chronicles* and J.R.R. Tolkien, author of the *Lord of the Rings* and *The Hobbit*. (He was called the father of modern fantasy literature.) Lewis wrote a great many scholarly books and articles as well as Christian books that I could almost understand.

George Washington, Thomas Jefferson, Benjamin Franklin, Abraham Lincoln. I wouldn't presume to identify the most admired American, but I assume that these gentlemen would be near the top of their list. Steve Jobs and Bill Gates are respected modern day geniuses.

William and Catherine Booth founded the Salvation Army to minister to the poor, not just by feeding them but also providing housing and developing ways to earn a living. Eventually they established the Salvation Army in 58 countries.

Florence Nightingale came from a wealthy family, but she studied to become a nurse. She overcame physicians' prejudices and poor sanitation procedures to reform military hospital practices, saving many lives and giving nursing the respect it deserved.

Helen Keller at one time was the most famous handicapped person in the world. Keller was blind and deaf but she learned to read and eventually attended college. She wrote a number of books and became an activist and lecturer in support of the blind and deaf. She also founded and promoted the American Foundation for the Blind and was regarded as one of America's most inspirational figures.

Winston Churchill was a giant intellect with great courage as he led the people of Great Britain (and essentially the rest of the free world) through World War II. Margaret Thatcher was another great British Prime Minister. William Wilberforce led the political campaign for over 30 years in 18th century England to make the slave trade illegal.

Edmund Hillary is also on my list, not just because he was the first to conquer and climb Mt. Everest, but also because he then spent much of his life helping the people of Nepal, for example by building hospitals and schools.

Queen Elizabeth could be mentioned, but I'm not a big fan of the Monarchy and its implied assumptions about the superiority and privileges of some people over others while accumulating massive wealth.

I have seen JFK, Oprah Winfrey and Bill Clinton on lists of most admired people. While they had great charisma and their abilities enabled them to accomplish many things, I have reservations about their personal integrity so I wouldn't put them on my list.

VIGNETTE 34

Is Anybody Hungry?

THERE HAS BEEN a lot of discussion in Canada about the rising cost of food, and this is a real concern for ordinary Canadians, especially those who barely make ends meet each month. Basic commodities such as wheat, corn and sugar are increasing in price, which in turn drives up the price of many other items such as cereals, dairy products and meat. Poor harvests regularly cause supply problems and subsequently increase the cost of food. And the cost of oil is always a wild card, relating to political instability in many countries.

Some experts project that in 40 years there will be at least 2 billion more mouths to feed, with the world population approaching 9 billion. We passed seven billion, apparently on October 31, 2011 and are projected to reach eight billion by 2025. In 1800 there were "only" one billion people, and it took until 1987 to reach five billion. Experts have been predicting for a long time that the population of the earth would overwhelm the food supply, but so far the dramatic increases in farm productivity have almost kept pace. A hundred years ago farmers used teams of horses (or oxen in some countries), while today modern North American farmers use tractors with the power of 40 to 300 horses. One North American farmer now supplies food for about 129 people, compared with just 25 people in 1960, and they work almost four times more land than their predecessors from 1900. Whether you support modern intensive farming practices that use fertilizers and pesticides, or organic farming practices, it's clear that, at least in Canada and the U.S., we can produce a lot of food. The capacity to grow food may still increase since scientists continue to work on strains of grains, for example, that will be drought or flood resistant.

And yet, people in many countries go hungry every day. A United Nations report indicated that 25,000 people die from starvation every day. A Red Cross report in 2011 indicated that over 1.5 billion people are overweight and there are more fat people in the world than ever before. Another report noted that that there are at least one billion people who are not getting enough to eat! Even in Canada too many people depend on food banks. How can this be allowed to continue? A major problem in developed countries is food waste. One report indicated that rich countries waste almost half the food that they produce. One study said that 25 percent of food produced goes straight to the garbage or is discarded by restaurants and grocery stores. Half of all salad, a third of all fruit and 20 percent of fruit and vegetables are thrown away uneaten. That comes to as much as 100 million tons of food that is just thrown away. Why can't we distribute a majority of this to hungry people? Because food has been fairly cheap and we can't be bothered? Food production techniques often gain the most attention, but so should food storage and distribution procedures.

In ancient biblical times, gleaning was practiced; this was meant for the poor to "glean" leftover or missed grain or other food plants. Landowners were mandated to ensure that sufficient unharvested food was left in the fields to enable the poor, the migrant and the orphaned to survive. In Abbotsford, BC, we have an organization called the Fraser Valley Gleaners (FVG), operating in much the same spirit. FVG is a non-profit, non-denominational Christian organization whose primary purpose is to make dried soup mix and apple snacks for distribution, in the name of Christ, to over 40 countries. This society produces mainly soup mix consisting of dried, fresh, and frozen Brussels sprouts, onions, tomatoes, carrots, peppers, beans, peas, broccoli, cauliflower, beets, turnips, potatoes and lentils. The vast majority of the donated foods are "seconds," but are perfectly good even though modern day "sophisticated" shoppers would turn up their noses at them in our fancy grocery stores and markets. This fastidious attitude forces grocers to throw away these "seconds," but many of them are willing to donate them to Gleaners. In winter FVG also dries the cuttings made available when grocers trim products (such as cauliflower and broccoli) in preparing them for the store shelves. Fresh and frozen vegetable produce and apples are donated by greenhouses, vegetable processing operations, and local growers. Volunteers to help prepare the soup mix are always welcome.

When water is added to the dried mix in each bag, 100 servings of nutritious soup are available to feed the hungry. Reliable, reputable distribution partners distribute these bags of hope throughout the developing world. In 2010, 10.5 million servings were processed and made ready for shipping.

Some people in our community also regularly go hungry, and Abbotsford's experience is also reflected in many other communities. During the recession of the 1980's many families and individuals were finding it increasingly difficult to provide for the basic necessities. In the early 1980's the Food Bank was assisting approximately 900 people every month. Today over 4000 people each month are helped, with approximately 40 percent of those being children, and this situation is found in many North American towns and cities.

VIGNETTE 35

Snoring vs. Leprosy

IN ANCIENT TIMES, people suffering from leprosy were treated as social pariahs. There was no understanding as to what caused the disease, which was wrongly believed to be highly contagious. Lepers were placed in separate colonies, with essentially no contact with regular people; food was left for them in designated areas to be picked up by the lepers after being left there by a family member. Actually, lepers have been treated pretty much like this up until relatively recent times, and many leper colonies reportedly remain around the world in countries such as India, China, Romania, Egypt, Nepal, Somalia and Liberia.

Snoring is a totally different kind of malady, and I don't mean to downplay the seriousness of leprosy. But is there anyone there among you who have not made fun of someone who snores? Have you not felt just a little bit superior to the poor soul whose misfortune it is to snore? Who among you has not used highly descriptive adjectives, with a considerable dollop of hyperbole and exaggeration, when describing the snoring characteristics of a loved one who snores? Snoring is one of those interesting conditions where one has to be told by someone else (usually enthusiastically, and typically by a person who does not snore), that they have this unfortunate, generally socially unacceptable problem tendency to regularly disrupt the sleep of someone else. My sweetie, bless her heart, has never ridiculed me for snoring. This, combined with putting up with my various other eccentricities should automatically qualify her for sainthood in the Catholic Church. Even the Mennonite Brethren Conference, which in its 500 plus years of existence has never officially designated someone as a "Saint", could perhaps make an exception in her case. But she has, with a certain exuberance, told more than a few people about my snoring prowess (no mention of the Richter Scale, but you get the idea), so perhaps official sainthood isn't a sure thing after all. Just because I'm not paranoid about this doesn't mean that many people out there aren't determined to mock me or other fellow sufferers.

Leprosy is also called "Hansen's Disease" and is caused by bacteria. It is a disease of the nerves and mucosa of the upper respiratory tract; skin lesions are the primary external sign. If untreated, leprosy can be progressive, causing permanent damage to the skin, nerves, limbs and eyes. It is believed that the bacteria causing the disease are usually spread from person to person in respiratory droplets. Dr. Paul Brand was a pioneer in developing tendon transfer techniques for use in the hands of those with leprosy. He was the first physician to appreciate that leprosy did not cause the rotting away of tissues, but that it was the loss of the sensation of pain which made sufferers susceptible to injury.

Statistics on snoring are often contradictory, but at least 30 percent of adults (frequency is higher in men than women) snore. There are a variety of factors that can lead to snoring, such as the anatomy of the mouth and sinuses, alcohol consumption, allergies, a cold, and body weight. Age can also be a factor; as we get older our throats can become narrower and we can start to lose muscle tone in our throats, which can lead to snoring problems. When a person dozes off and progresses from light sleep to a deep sleep, the muscles in the roof of the mouth (soft palate), tongue and throat relax. The tissues in the throat can relax enough that they vibrate and may partially obstruct the airway. And, the more narrowed the airway, the more forceful the airflow becomes. This causes tissue vibration to increase, which makes your snoring grow even louder.

Besides the noise of snoring, more complex conditions such as sleep apnea can be consistent with the symptom of snoring. Snoring may be irritating or sometimes funny or quirky, but it can also be serious if the throat tissues obstruct the airway, preventing the sleeper from breathing. Sleep apnea is often characterized by loud snoring followed by periods of silence when breathing stops or nearly stops, since the airway becomes so small that the airflow is inadequate. Eventually, the lack of oxygen and an increase in carbon dioxide signal the person to wake up, forcing the airway open with a loud snort or gasping sound.

Popular treatments for snoring proposed on TV are generally marginally effective, if at all. The Band-Aid strips which are supposed to widen the nasal air passage may be effective on TV and with a few folks, but not for me. I have been informed that ear plugs do not work, unless you are talking about the super-strength ones that are used on the deck of an aircraft carrier. The use of pillows judiciously placed over the head of the snorer has been proposed, but this has certain serious legal ramifications. Going pillow free has also been recommended, and conversely elevating the person's head has been recommended to facilitate breathing. Another suggestion is the "tennis ball trick": sleep with a tennis ball in a sock sewed to the back of the pajama top. The tennis ball is uncomfortable if you lie on your back, and so you will likely respond in your sleep by turning on your side. Soon you will develop side-sleeping as a habit and not need the tennis ball.

Surgery, which unfortunately is considered by the government as cosmetic or elective surgery (thereby demonstrating a callous indifference to the emotional trauma experienced by snorers or their spouses), so that only the very desperate or those who also have extra cash in their jeans, take advantage of this option. Though snoring is often considered a minor affliction, snorers can sometimes suffer severe impairment of lifestyle. One clinical trial discovered a statistically significant improvement in marital relations after snoring was surgically corrected.

VIGNETTE 36

We've Become Our Parents

SOMEHOW PEOPLE, ESPECIALLY my kids, have gotten it into their heads that I am not cool. I know, I know, I'm sure you are as shocked as I was. I have been known to be cool, some years ago (well, perhaps a lot of years ago), so what changed? Gray hair and the loss of hair for one thing, which don't give the cool image that we see in many TV commercials. And, on top of losing my coolness, I apparently have become an old codger, that is to say, I have gotten old. I never dreamed that I would ever become as old as my parents.

It seems no matter how much we might love our parents when we are kids, whether we're 10 or 25, we think that Mom and Dad are decidedly old fashioned, and gosh, I hope I never look or act like they do. I remember visiting my folks each year (they lived in a different city) and thinking, "Wow, the TV is loud enough that the neighbors probably know all the programs that Mom and Dad are watching every day." Well, guess what, my kids now say, "Do you really need to have the TV that loud?" For another thing, computers and technology have exacerbated my loss of coolness. I seem inherently unable to understand either the terminology or what procedures to follow to make simple things (simple for cool people) like DVD players or how TVs less than 15 years old work.

I confess to thinking that my parents dressed pretty much like old farmers that hadn't ever thought about fashion. Well, the more things change, the more they stay the same! We recently went through all of the pictures that we have taken over the past many years, and the grandchildren laughed like crazy at their own parents who had laughed at us! "How could you dress us so funny?" our kids said, but I told the grandchildren not to laugh too much, since even though they are very cool now, the time will come when their kids will laugh at them for their perceived lack of fashion sense.

Can you repeat that? Pardon me??? I confess that my hearing isn't as good as it used to be, and I often have to ask my wife to repeat what she said, a lot just like my Mom used to do ... Sorry, I must have nodded off for a minute. This happens sometimes if I'm bored, but lately it seems to happen more often. My doctor is recommending a hearing aid, or at least that's what I think I heard her say. And the prescription for my glasses seems to be getting stronger. Of course, that's after the cataract surgery that I had on each eye two years ago. The annual medicals I had at work every year were just an opportunity to visit with the nurse back then, but recently I have had to visit my doctor more than once a year. At least she's not on speed dial yet. She always runs her fingers through my hair, but my sweetie tells me that's only because I have these funny bumps on my head that need medical attention.

I used to think that my folks kept their home at a hothouse temperature. Guess what? When I visit my daughter and son-in-law's house I wear a jacket and make sure that I wear my slippers, because they keep their house so cold, and they laugh! I told them that their thermostat probably is set at 20 degrees Fahrenheit rather than Centigrade, but they just laughed harder.

As my dad's reaction time slowed and his eyesight deteriorated, my mom used to help him drive –"The light is turning red; you need to turn right at the next corner; look out for that car in the other lane; and I think I hear a siren so pull over when you can." Well, now my wife is giving me the same instructions, and, although my memory isn't so great any more, I think she is doing this with increasing frequency and urgency.

I used to be impatient when my parents loved to stay home. Now a night at home seems attractive. I can remember driving to work from one suburb to another early in the morning on those rare occasions when there had been a foot of snow overnight. I thought this was a great adventure, and the toughest decision was deciding which steep hill to take off of Marine Drive in Vancouver to end up at my destination. Now, if I see more than 17 snowflakes come down I either rush home or decide the weather is much too severe to venture out.

How about bus tours? Only old people, like my parents, used to go on bus tours. My wife and I were young and adventuresome and used to drive across the country in summer, but now even we have gone on a bus tour.

Growing old is not for sissies. When I first saw this on a T-shirt I laughed. But then I thought about it, and it's very true. It takes a tough old bird to get through this aging thing. I know they say growing old is a state of mind, but that's what the young say. They haven't really experienced the aging process yet and they are concentrating on being cool. It isn't cool to think about our bones gradually deteriorating throughout our life by a process of absorption and formation, and as we age the balance changes, resulting in a loss of bone tissue. What young person thinks about their bones becoming less dense and more fragile as the mineral content of bones decreases, leading to possible osteoporosis and the probability of broken bones when they fall? Well, uncool old people do! Many old folks wonder if their osteoporosis will change their spine and cause them to develop a "dowager's hump." But I differed from my parents since I realized that many of the changes in our musculoskeletal system result more from disuse than from simple aging. While only 10 percent of seniors participate in regular exercise, I decided to be different by getting at least moderate amounts of physical activity to reduce the risk of developing high blood pressure, heart disease, and some forms of cancer. Now, maybe I have a chance to reach the proverbial "three-score and ten" since I bike quite a bit, I hike sometimes or take walks with my sweetie, and I play tennis periodically with the grandchildren. But I am realistic enough to know that I'll never be cool again.

"If you live to be one hundred, you've got it made. Very few people die past that age."—George Burns

VIGNETTE 37

Worship Music in Church

LIKE MOST PEOPLE who go to church, I tended to think of worship as something that moves our thoughts and desires from here on earth to God the Father and Jesus the Son in heaven through the power of the Holy Spirit. We often think of worship as something that originates with us, our gift to God, but perhaps that's why so many of us are conflicted about it. We consider worship to be an expression of our personal devotion, and to some small extent this is true, but our personal devotion is often lacking.

So when the music style or service format or worship expression is different from what we want or feel, it gets in the way, and we either don't feel like it is worship and/or we criticize what we see happening. For John Koessler, a theology professor, the Biblical portrayal of worship moves in the opposite direction—the "trajectory" of worship begins with God and descends to earth. Psalm 150 is a good example, where praise begins in the heavenly sanctuary and resounds throughout heaven and earth. Koessler says that worship is not our attempt to project our voices so that they will be heard in heaven, nor is it a performance executed on an earthly stage for the benefit of the spectator, God. And we certainly should not worship or lead worship to make ourselves feel good. Ideally the worshipping church doesn't merely imitate what goes on in heaven, but we also participate in heaven's worship. I have also heard people say that they "didn't get much out of the worship service today," when the better thought would have been, "I hope that God was pleased with my worship today."

Psalm 150 verses 3–5 says "Praise Him with trumpet sound; praise Him with lute and harp! Praise Him with timbrel and dance; praise Him with strings and pipe! Praise Him with sounding cymbals; praise Him with loud clashing cymbals!" Almost everyone agrees with these verses, but where we often, very often it seems, disagree is the kind of music we should have in church and which instruments should be used in our worship. As Charley Brown would say, "Good grief!" Could strings mean guitars and loud clashing cymbals be drums!? And why isn't the organ or piano mentioned!? We grow accustomed to certain instruments and prefer particular styles, and in the process of not wanting to change we lose sight of why we are worshipping.

Of course we will be profoundly affected by the music we hear, and some of it we won't like. Koessler goes on to say that the Psalmist's description of music suggests that the variety of music styles, the instruments used, and the methods the church uses in worship should exceed the scope of our taste. So let's agree that we can't please everyone, and that the quality of music is not the most important issue in our worship experience. More importantly, it's not our differences in taste but rather our lack of respect for others that causes much of the damage in the church regarding music and worship. We don't seem to be able to sacrifice some of our preferences so as to be able to accommodate others.

Praising the Lord should be the outflow of a person enjoying the provisions of God; the outflow of a full heart. Let's remind ourselves for now that we are part of a much larger congregation of patriarchs and prophets, saints and angels, and we are invited to join a chorus that began at creation.

Some hymns are fabulous and have great theological or uplifting messages and have been sung for hundreds of years. Others, not so much. Some choruses, "sung off the wall," are also solidly theological and uplifting, others not so much. Many believe that contemporary songs are more "seeker friendly." One response is that the desire to reach out to non-Christians is wonderful, but that doesn't mean that this approach is necessarily correct. Accommodating the uncommitted individual's consumer preferences or implying that, "We're just worshipping God here and have no standards other than pleasing you," is not the best way to do evangelism. It may be better to say, "We do sing some old songs and hymns, and lots of contemporary tunes, and that's because in each case they reflect our views of what God is doing for us through Christ our Saviour." The sincere seeker may think, "Here's a church that knows what it believes and is willing to stand by their convictions."

VIGNETTE 38

What Is a Split Infinitive, and Why Would Anyone Want to Dangle a Participle?

I TOOK ONLY one English course in university, since I was a Pharmacy major, and English 101 was the one class we dared to skip periodically, so please don't expect superb grammar in this little dissertation. People's ability to write varies widely and it's a real treat to read an accomplished author, both in terms of seeing excellent grammar and to also see their wide-ranging and rich vocabulary.

Grammar is the set of structural rules that govern the composition of sentences, phrases, and use of words. The term refers also to the study of such rules, and this field includes morphology and syntax, often complemented by phonetics and semantics. Usage books and style guides that call themselves grammars may also refer to spelling and punctuation. Computer grammar and spell checkers help sometimes.

Personally, I could never remember if quotation marks should come before or after the period. It turns out that there is a difference between U.S. and British/Canadian punctuation styles. In the U.S., periods and commas always appear inside the quotation marks, as in "Let's go to the zoo." For British/Canadian punctuation the punctuation follows this logic: If you are quoting a question then the "?" will go within the quotation marks. For example: Sally said, "Where are you going?" But if you're asking a question about a quote, then the "?" will go after the quotation marks: Did Sally say, "We are going to the zoo"? If you forget about this second example, you'll be right more often than not.

Split infinitives happen when you put an adverb between *to* and a verb, for example: "She used to **secretly** admire him", or "You have to **really** watch him". What's wrong with this? Some people believe that split infinitives are grammatically incorrect and should be avoided at all costs. They would rewrite these sentences as: She used **secretly** to admire him, or, you **really** have to watch him. But people have been splitting infinitives for centuries, especially in spoken English, and avoiding a split infinitive can sound clumsy. It can also change the emphasis of what's being said. For example, you really have to watch him, as in 'It's important that you watch him' doesn't have quite the same meaning as: You have to really watch him, that is 'You have to watch him very closely'. To split or not to split? The 'rule' against splitting infinitives isn't followed as strictly today as it used to be, but it's safest to avoid split infinitives in formal writing unless the alternative wording seems very clumsy.

A participle is a word or phrase that modifies the subject of a sentence. Usually the participle is the present or past tense of a verb, and can also be used as an adjective. Dangling participles are tricky words or phrases that change the meaning of a sentence so that it isn't clear exactly what we mean. Because of where a participle is placed in a sentence, a verb that was intended to modify the subject of a sentence confusingly seems to modify the object.

Some examples may help: "Driving home in yesterday's storm, a tree fell on the back of my car;" or, "Crossing the room, her foot bled all over the carpet." I hope that you can see the problems with these dangling participles! How about another ridiculous phrase? "I saw the house peeking through the trees."

But most of us have long forgotten basics like adverbs, which describe verbs (action words that tell us something about the subject of the sentence). In "Giselle will plant twenty tulip bulbs next week," the compound verb "will plant" describes an action that will take place in the future. A preposition links nouns, pronouns and phrases to other words in a sentence. The word or phrase that the preposition introduces is called the object of the preposition. There seems to be confusion about whether it's poor taste to end a sentence with a preposition. I leave it to Winston Churchill to shed some light on this: "Ending a sentence with a preposition is something with which I will not put." Class dismissed!

VIGNETTE 39

Airbags and Other Interesting Chemical Reactions in Automobiles

I NEVER REALLY stopped to think about how airbags worked until some of our family members were saved from serious injury when the airbags on their van deployed. Airbags were first developed for airplanes as crash landing devices in the 1940s, and were also used as crash landing devices for spacecraft, but they hadn't been widely used in passenger vehicles until fairly recently.

Airbags work using several chemical reactions. The undeployed airbag is folded inside the car inside the steering wheel, or above the glove compartment, and now also along the side doors. There is a sensor at the front of the car that measures acceleration, or in crash cases, rapid deceleration, and when a vehicle stops very suddenly this triggers the inflator connected to the airbag, which in turn produces an electric spark that deploys or activates the airbag. Inside the airbag is a stable compound called sodium azide, and when this compound is ignited the resulting explosion releases a large volume of non-toxic nitrogen gas through the decomposition of the sodium azide pellets. This reaction also produces sodium metal, which is unstable and potentially explosive, but the sodium quickly reacts with a second compound, namely potassium nitrate, which helpfully also generates additional nitrogen. There is one more reaction involving potassium oxide and sodium oxide (by-products of the second reaction), resulting in the formation of harmless silica. A handful of sodium azide produces about 70 liters of nitrogen gas, which is enough to inflate a normal airbag in about 1/3 of a second, that is, quickly enough to protect the passengers.

Once the airbag is deployed, deflation begins almost immediately as the gas escapes through vents to soften the impact of the car occupant as they surge forward. Dust-like particles of corn starch or talcum powder that lubricate the airbag during deployment are also released, which explains the powdery substance that you see after the airbags have been deployed.

Disposal of vehicles containing undeployed airbags in landfills must be done with care, since the sodium azide is an environmental hazard, plus the airbags potentially could cause an explosion. Authorities generally require airbags to be physically deployed before vehicle disposal to eliminate this potential hazard.

Chemical reactions such as those just described that occur in vehicles are not just helpful in providing safety. Cars and trucks of course turn fuel into energy through a series of chemical reactions. Gasoline, or diesel fuel, consists mainly of medium-sized hydrocarbons. Inside the piston chambers of an engine, the fuel is sprayed into the cylinder to mix it with the air's oxygen and this mixture is then ignited. When the oxygen-fuel mixture is ignited the combustion of the hydrocarbons releases a lot of energy in the form of heat. The reaction, which is basically a small controlled explosion, produces heat in the cylinder, making the air expand and pushing the piston outward. The hydrocarbon molecules are changed into carbon dioxide, carbon monoxide, and water as well as some other by-products. A car engine uses the energy produced by this reaction to push its pistons up and down and this is how the engine converts the gasoline's chemical energy into mechanical force. Basically, the force from the expanding gas in each cylinder ends up putting torque on the axle so that the vehicle is driven forward.

Some people are looking to biofuel as an alternative to power their cars since the price of gasoline keeps increasing. It's possible to make your own biofuel at home using some basic chemistry if you are both creative and have some practical skills (for a cost often much cheaper than what you pay at the pump), using vegetable oil, methyl alcohol and lye. The lye and methyl alcohol produce sodium methoxide, which removes the thick glycerin from the vegetable oil. This chemical process, which can be done in your garage, requires a well-ventilated area at temperatures above 70 degrees F (21 C). The biofuel that is produced can be used in modern diesel engines. It is recommended that at temperatures below 55 degrees F (13 C) the biofuel should be mixed with 50 percent petroleum diesel, and for older engines the ratio should be 20 percent biofuel and 80 percent petroleum diesel. The chemical reactions in converting chemical energy into mechanical energy are of course basically the same.

VIGNETTE 40

Is that You Glowing in the Dark?

THE SHORT ANSWER is, "No." After the 2011 earthquake in Japan, I read a column that said "radiation, if I may use the pun, is back on the local radar." People should have a healthy respect for radiation, but instead we usually see an inordinate fear. We forget that we are all exposed to ionizing radiation every day, and that we have sophisticated monitors in place to alert us if the background levels increase. A "Sievert" measures the biological effect of absorbed radiation, but the "millisievert," which is one-thousandth of that dose, is the common unit to measure exposure. Experts tell us that we are exposed to less than 3 millisieverts per year. A few weeks after the Japanese earthquake, monitors in North America recorded just .0005 millisieverts from Japanese sources. Even the human body is radioactive—just by sleeping next to another human body we are exposed to about that level of radiation.

Some would argue that there is no safe level of exposure to radiation, but that's misleading. That's like saying that sodium chloride (everyday salt) is hazardous (and it is toxic at high doses). We determine what a safe level is, do a risk to benefit assessment, and if we are below that level we continue to live our lives. We would need to walk through a thousand airport security scans to equal one chest X-ray, but we do have an x-ray when it is necessary. A full body CT scan may expose us to about 30 millisieverts, but we decide that the risk to benefit assessment makes this worthwhile when a health issue needs diagnosis. On one cross-country flight we absorb about .04 millisieverts. An airline pilot absorbs about

2.2 millisieverts per year. Health Canada says that naturally occurring radiation provides about 89 percent of Canadians' average annual exposure to radiation. They also set the safe maximum annual exposure for nuclear workers at 50 millisieverts, which is 100,000 times the level of radiation measured here on the West Coast at the end of March 2011.

What is radioactivity? It's difficult to give an answer without plunging into technical terms and details. Basically, it's the spontaneous emission of energy, either directly from unstable atomic nuclei or as a consequence of a nuclear reaction. A radioactive substance emits radiation, and includes particles such as electrons (beta particles) and gamma particles in an effort to become more stable. Some elements are much more stable than others. Plutonium and uranium are common examples, but familiar elements like carbon are "radioactive" as well. Radioactive decay is the process by which an unstable atomic nucleus loses energy by emitting ionizing particles (ionizing radiation). Uranium, with a molecular weight of 238, is a very unstable element, and it goes through 18 stages of decay before becoming the stable isotope lead 206. An unstable nucleus like uranium can be split up into two other nuclei through nuclear fission, releasing tremendous amount of energy in the process. This is the principle on which nuclear energy and nuclear weapons are based!

Half-life is the amount of time required for half of the radioactive element to decay. Carbon usually exists as C-12, which is much more stable than C-14. For example C^{14} has a half-life of 5730 years. That is, if you take 1 gram of C^{14}, then half of it will have been decayed in 5,730 years and half again in another 5,730 years. Iodine ^{131}, on the other hand, has a half-life of 8 days.

Radiation exposure may cause cancer if it causes a loss of control of cell division, and cells begin dividing uncontrollably. On the other hand, radioactivity can also be used to treat cancer since the high-energy radiation kills rapidly growing cancer cells and shrinks tumors. Radiation therapy kills cancer cells by damaging their DNA (the molecules inside cells that carry genetic information) so that the cancer cells stop dividing or die. The radiation may be delivered by a machine outside the body, or it may come from radioactive material, like iodine^{131} placed in the body near cancer cells.

As the world's population increases, and there is continued desire to improve living standards, there is likely to be demand for more electricity, and nuclear power should be considered. Energy sources available in the world include coal, nuclear, hydroelectric, gas, wind, and solar. Every form of energy generation has advantages and disadvantages. Coal is inexpensive and generally easy to recover from the earth, but it requires expensive air pollution controls and is a contributor to acid rain and global warming. One report suggested that burning coal to produce electricity is a greater threat to health than the radiation released by the 3 worst nuclear accidents combined (Three Mile Island in the U.S., Chernobyl in Ukraine, and Japan). For nuclear power, the fuel is inexpensive, the energy generation is more compact than other sources, but it requires larger capital cost because of emergency procedures, containment precautions, expensive radioactive waste and storage systems, plus dealing with accidents. Hydroelectric power is inexpensive once expensive dams are built, and locations depend on water elevation, but dams can cause environmental damage either by flooding valleys or by collapse. Wind is free, but it's not efficient and it's limited to windy areas. Solar power is also inexpensive when sunlight is available and costs are dropping, but current technology requires large amounts of land for the small amounts of energy generated, and special materials are required.

VIGNETTE 41

From Smoke Signals to Cell Phones to Twitter to....?

HAVE YOU EVER been asked by a child, "Were you ever chased by a dinosaur when you were small?" Well, I haven't either, but, even worse, I feel like a dinosaur. I can remember, many years ago, watching a detective movie on TV where the investigator discretely used a tape recorder to record the telephone number a suspect was dialling on the public dial phone. I thought that was pretty ingenious. I can also remember the party line telephones in rural Alberta where one would hear several clicks as neighbors picked up their telephone receiver to listen in on your call. And I recall visiting the Alexander Graham Bell museum in Nova Scotia where the telephone was first invented. (Mr. Bell's first words on his new phone probably were to his teen age daughter, "Please get off the telephone so I can see if this thing works"!) I used to read stories about the building of the railway in Canada and the use of Morse Code and its system of dots and dashes, which some could instantly decode the telegraph messages. Now Morse Code is pretty much dead since it has been superseded by more modern means of communication.

I was pretty proud of myself when I got a cell phone a few years ago, but, being pretty much of a Luddite, I only use it for phone calls. Selecting a unique ring tone was something special back then, and, no, I had enough social awareness not to select "The Yellow Rose of Texas". An increasing number of households no longer have land lines since they rely totally on smart phones and iPods and things that are stuck in their ear. I also have a GPS since I am directionally challenged, but I hear that the stand-alone GPS will soon be obsolete since the new cell phones will do the job, thank you very much. Plus they will take better pictures than I ever could even with a nice camera. But I read in the papers that making phone calls at all is now regarded, by those who apparently know, as uncool. A guiding social principle used to be "Don't call anyone after 10 p.m.," but now some really cool people like celebrities and other upwardly mobile people are saying, "Don't call anyone. Ever." The correct social interaction now apparently is texting. While this may be somewhat extreme, for now at least, it seems that traditional phones may be ringing much less often in the future.

Is this a good thing or not? I have seen teens and twenty-somethings sitting side by side texting each other. What ever happened to conversation? Is sending text message a solid way to build a friendship or a romance? Doesn't anyone miss meandering telephone conversations when people had more time (or made more time) for each other? Apparently, even sending e-mails is becoming a burden since they are much too time consuming. Here I thought that electronic mail was a great thing since one could, almost instantly, convey thoughts and impressions and details to friends and business contacts and family members without wondering if the snail mail would lose my letter or reroute it via Mexico City. A few years ago, keeping in touch while travelling abroad was easy and satisfactory if you could find a sympathetic hotel clerk. Then we looked for an Internet Café, but now we use our smart phone or laptop. Status updates have now been replaced by texting and Twitter and everyone's pictures are to be found on Facebook.

It seems that we humans have an inexorable need to communicate with each other. There's anecdotal evidence that our need to be connected "24//7" is as addictive as drugs can be. Some companies require employees to put their BlackBerrys in a box before they enter meeting rooms. Some students in universities or high schools forced themselves to change their friends' Facebook passwords so they couldn't log on during an exam and procrastinate rather than answer the exam questions. How many times per day do you check your e-mail or catch up on Facebook or Twitter? Fifty and more years ago, people incessantly wrote letters, sometimes several times per day; this wasn't instant communication but

From Smoke Signals to Cell Phones to Twitter to....?

it satisfied their need to stay in touch. It's interesting to speculate what the next steps in instant and constant communication will be. There's a line in the old musical "Oklahoma" that describes the modern status of Kansas City when it says, "They've gone about as far as they can go." I suspect that this statement doesn't apply to human communication.

VIGNETTE 42

That's the High Price of Being Canadian, Eh?

WE MODERN DAY folks must be getting stronger. I can remember that it took my Dad and four kids to carry $50 worth of groceries, and now I can easily do this with one hand. Or could it be that groceries, and everything else, are much more expensive today?

Airports and airfares are an example of the high cost of living in Canada. Over one million travellers drove to Bellingham or Seattle in 2010 from BC because of the substantially higher costs of flights originating in Canada. A million people would fill an additional 25 flights per day if these folks left from Vancouver or Abbotsford. Wouldn't you think that it would be an economic stimulus rather than a total loss of business if at least some of these flights could originate in Canada? A Vancouver Sun article in 2011 said that taxes and airport fees are six-fold what they are in the United States.

Allegiant's lowest fare from Bellingham to Las Vegas was $99, or $110 with taxes and fees. WestJet's lowest fare from Vancouver to Las Vegas was also $99 but became $169 when, you guessed it, taxes and fees are included. U.S. airports don't have airport improvement fees, and they don't have NavCanada fees or GST or HST, and they have lower rents. For a long time, airport improvement fees were collected at YVR (they still are, but now they are hidden), but the federal government still collects large amounts of money from airports. I'm not sure how much you can blame the airlines, but we sure need to look at airports

That's the High Price of Being Canadian, Eh?

and governments. A WestJet fare of $39 from Toronto to Montreal becomes $100 with taxes and fees. A $39 fare from Boston to New York becomes $51 with taxes and fees. Parking at YVR costs $13 per day, plus taxes, Abbotsford is $8 per day plus tax, and Bellingham is $9 per day including tax. Apparently, 78 airlines have failed in Canada in our history, and it doesn't take a rocket scientist to figure out that high operating costs were probably a big factor. Disney is running cruises out of Vancouver in 2010 but will do this out of Seattle next year due to the cost of airfares to Canada.

And this idea of massive taxation isn't limited to flying. Metro Vancouver has a 21 percent (**Twenty-one per cent!**) parking sales tax, I presume in addition to other taxes like HST or PST, so parking at the Vancouver Airport for a week comes to a huge total. For a long time, books have had two pricing labels, one for the U.S. and one for Canada, and the price difference bears no resemblance to the exchange rate. Canadians pay 14 percent more for books, 28 percent more for Blu-Ray, 15 percent more for clothes and 24 percent more for the iPod Touch. The price of gas for our cars and trucks is another example where the U.S. cost is significantly less. BMO Canada issued a report in April 2011 that said we pay an average of 20 percent more for everything from movies to running shoes, while the Consumer Association of Canada estimated that we pay 30 percent more for things like TVs and cars than our American counterparts. This price difference is seen in many other products, partly because Canadians just grin and pay it, eh! Part of this price discrepancy is because the American market for consumer goods is more competitive and retailers try very hard to keep prices in check, while Canada's smaller population doesn't merit the same consideration. But Canadian retailers best try harder or Canadians will put up with longer line-ups at the border, especially when the Canadian dollar is close to being at par or better with the U.S. greenback.

It needs to be said that comparisons with the U.S can be problematic, and there are many examples of where we clearly do not want to be like them. Health Care comes to mind. But there are a few things we can learn from our big neighbor. We trail the U.S. significantly in productivity, that is, the measure of how much is produced for each hour worked. Somehow this continues to be the case even though we seem to have handled the recent economic downturn better. Chronic uncertainty about the threat of a labor shutdown at Port Metro Vancouver is said to be a major growth limiting factor at this port. There is another factor to keep in mind; the U.S. market has substantial economies of scale since they have at least 330 million people while Canada's population is about 34 million. But I think we have a kinder, gentler society, even if we have higher taxes in some areas.

Oh, Canada!

VIGNETTE 43

Ten Interesting Cities I Have Visited

ST. JOHN'S, NEWFOUNDLAND-LABRADOR. Putting this little city of about 200,000 on the list may be surprising. I was there one November, so winter hadn't quite set in yet, but the steep mazes of streets with brightly colored houses often perched on rocky out-crops, give the oldest English founded city in North America an attractive and distinctive architectural look and feeling. Water Street, winding above the harbor but along the downtown center with old stores and restaurants, is almost 600 years old. Marconi received the first transatlantic wireless signal in St. John's on December 1901 at Signal Hill on the edge of the city. Just to add a different touch, policing services for the city are provided by the Royal Newfoundland Constabulary.

Venice, Italy. The colorful houses, the canals winding their way throughout the city, the beautiful city square, the hundred or more islands that constitute the city, the tour boats jostling for "parking" or docking spaces, the delightful "hole-in-the-wall" restaurants, and the gondoliers ready to give tourists a ride all contribute to the scintillating atmosphere of a vibrant city.

Paris, France. We had heard that the city of lights, while beautiful, was difficult to visit since Parisians were arrogant and impatient with tourists. But we enjoyed our time there and all of the major highlights including The Louvre, going up the Eiffel Tower plus seeing it sparkle at night, the Arc d'Triomphe, the "Left Bank", the Avenue des Champs-Elysees, Notre Dame Cathedral, plus the restaurants and shops. We needed more time!

New Orleans, Louisiana. I visited there pre-Katrina days when it still had over a million people and grand memories of days gone by with beautiful old buildings. And of course there was the bustling French Quarter along Canal Street with jazz bands on the balconies in the evening. The walkway along the Mississippi and the hot, humid temperatures were memorable.

London, England. There's so much history and so many attractions to enjoy ranging from the relatively new London Eye to Big Ben and the Houses of Parliament, Buckingham Palace, Harrods, the Piccadilly Circus plaza, and more museums that you can count. I was fascinated by the underground Cabinet War Rooms that Churchill used during WWII, and I always like getting around in the "Underground" or "The Tube".

Lucerne, Switzerland. The Swiss are very neat and organized and Lucerne, with its stately buildings, shops and churches reflects this. The surrounding mountains with the snow-capped Swiss Alps in the distance and Lake Lucerne add to the magic.

Rome, Italy. History is a dominant feature in Rome; as you drive around you can periodically see ruins that may be thousands of years old, but you also see historic places like the Colosseum, the Forum, Circus Maximus where chariot races were held, and of course Vatican City with the Sistine Chapel and St. Peter's Basilica. We didn't have pizza while we were there, but we did have some fabulous meals.

New York City, New York. Actually we only visited Manhattan and didn't see the other four boroughs of Brooklyn, Queens, Staten Island and The Bronx. We took two one day hop on hop off bus tours to see the Empire State Building (and a trip to the top), Times Square and Broadway, Greenwich Village where we had a smoked meat sandwich, the United Nations buildings, the Brooklyn Bridge, the contrasts of Harlem and Fifth Avenue, plus St. Patrick's Cathedral. We had an enjoyable walk through Central Park; we were brave because we were with a local resident. We came by train to Penn Station way down below Madison Square Garden.

Tallinn, Estonia. We stayed in a Holiday Inn, which sounds pretty ordinary, but right across the street was the historic and impressive old walled, cobblestoned city that goes back a thousand years or so. Walking up and down the winding, hilly streets was always an adventure with many colored wall-to-wall buildings and domed massive churches just around the next corner. The regular city with miles of nondescript three or four story apartments along the street car line was rather drab, bringing to mind the influence of the then recently pushed out Russian influence.

Vienna, Austria. Every city has a personality, and Vienna's seems to be royal and cultural. We attended a Mozart-Strauss Concert in a classical, beautifully appointed hall, we saw the influences of past kings and queens and we saw the classical buildings and cathedrals.

Abbotsford, BC. You may think this is an odd extra inclusion, given the magnificent cities mentioned above, but my home town has some pretty attractive features as well. Called the "City in the Country," Abbotsford has a population approaching 150,000 with parks and walking trails and more shopping than you need. There must be more than a dozen streets from which one can see the Sumas Prairie linking to the majestic Mt. Baker just across the border in Washington State. Abbotsford's area is 359 square miles, which means that the city limits include many pastoral blueberry and raspberry farms, plus dairy and chicken farms, in addition to the features of a modern, vibrant city.

VIGNETTE 44

Buy Land. They Ain't Making Any More of the Stuff

THAT'S WHAT WILL Rogers, America's homespun philosopher, said many years ago.

As farmland dwindles in BC's most fertile regions such as the Lower Mainland adjacent to Washington State, hemmed in by mountains, forests and the ocean, questions are being raised as to whether agriculture even makes economic sense anymore. Many believe that the culprit is a steady erosion of government support of farmers, combined of course with incessant pressure from developers. The solution needs to involve a formal government policy defining food as a basic human right, and guaranteeing Canadians access to an affordable, local supply of food, along with policies to support such a declaration. Do you want to depend on getting strawberries and blueberries from California or Mexico? What happens when Ecuador decides that they can't export any more potatoes or snow peas, or when Chile no longer exports grapes or apples or beans, and we don't have enough farmland to grow our own?

All across the Lower Mainland it's the same story: farmers trying to squeeze a profit from islands of fertile land in a sea of urban development are fighting a losing battle. Farmland is shrinking and farmers are an increasingly rare breed—the average age of Canadian farmers is now about 52 years and increasing each year. In fact, all across BC agriculture is on life support, propped up to some extent by the provincial Agricultural Land Commission, which is mandated to stem the erosion of farmland,

and by federal and provincial subsidies and marketing boards. Agriculture isn't just another industry that should be left to the implacable forces of supply and demand. Many in the industry also believe that farming can't just be left to the market. The numbers leave no doubt: left to market forces, agriculture in BC would almost disappear in a few years, at least in the Lower Mainland and the Okanagan region of BC where the cost of land is so high. According to "BC Business" the average farm in BC recorded net income of just $26,213 in 2008. And almost half of that came not from the sale of crops and food products but from government programs and insurance payments. Those kinds of returns don't provide much incentive to shoulder economic risk and toil long hours at the mercy of unpredictable weather and markets.

BC's Agricultural Land Commission was legislated into existence in 1973 with a mandate to hold the line on shrinkage of the province's farmland—and to its credit, it has succeeded to some extent. Since then, the total area of farmland across the province has held steady at almost 5 million hectares, or about three per cent of the province's land mass. But this is misleading—the numbers reveal an inexorable shift away from the fertile Fraser Valley and temperate regions of the south, and toward the less-populous north. Since 1973 the south coast region, including the Lower Mainland and Fraser Valley, has lost 14,329 hectares of farmland. The Okanagan, Kootenay and Interior regions have lost 44,075 hectares. Meanwhile, the northern half of the province has gained 81,425 hectares as reported by BC Business.

Farmers are faced with tough decisions. Many dairy farmers have left for greener pastures; many couldn't afford to sit on property worth perhaps $100,000 an acre, or more. Many of them sold their farms, and bought properties more than twice the size in the BC Interior or Saskatchewan or Alberta, and pocketed enough change to finance a comfortable retirement.

Some farmers bought dairy farms which were going out of business around Abbotsford a few years ago and planted blueberries when a consumer craving for the berry pushed wholesale prices to near records. Today a farmer may gain much less, and in some cases may wonder if it is worthwhile to harvest the berries. Now the poor guys who bought property at $90 or $100 thousand an acre may not even be breaking even on expenses, never mind mortgage payments, if the price of their crops drops too low.

Urban sprawl is a major pressure. There's a saying that blacktop is the last crop; once farmland is paved over, nobody's farming it. Land is often divided up for building estate houses and supposed hobby farmers who often don't do any farming. Some say this is a minor irritant compared to the irrevocable loss of land to development. Developers, eager to make huge profits, are an even bigger challenge. Delta BC, a southern suburb of Vancouver, is a good example where numerous zoning battles have seen close to 400 hectares of farmland threatened by development in recent years; a First Nations treaty took 182 hectares out of the Agricultural Land Reserve; 70 hectares were lost to a rail yard expansion adjacent to the Delta port; 90 hectares lost to the South Fraser Perimeter Road; 56 hectares ceded to expansion and development of a Golf Club in the town center of Tsawwassen.

Whether our shrinking farm base is an economic catastrophe depends on which side of the fence you're on. To agrarian activists it's a question of national identity; about things like "food sovereignty" and "the right of a nation to chart its own course with respect to food." Do we want to be dependent on importing most of our food and become dependent on the vagaries of political instability and market competition? To all those dairy farmers who cashed in their valuable Fraser Valley real estate, it's simply a matter of going where the money is for their families, but there are bigger societal issues to also consider.

VIGNETTE 45

Urban Farming—Look No Further than Your Own Backyard

THERE'S A SOMEWHAT new approach that may represent the future of growing some of our own food, at least in cities, and it doesn't involve greenhouses. And especially it doesn't involve building a laneway house, which was one of Vancouver's recent attempts to increase city density and provide family affordable housing. No, it involves entrepreneurs and ordinary folks who want to use backyards and other available areas to plant gardens and grow fruits and vegetables.

One such entrepreneur cultivated the backyards of 13 of his East Vancouver neighbors, and his customers paid an annual fee of several hundred dollars for a share in the pooled yield. Every week throughout the growing season, clients bring their baskets to a central location and help themselves to their portion of the harvest: spinach, lettuce and rhubarb in the spring; carrots, beans and tomatoes throughout the summer; beets, broccoli and Brussels sprouts in the waning days of August and into the fall.

Rather than working for a lawn-maintenance company, one innovator thought that he could probably make more money on a 10-by-four-foot raised bed of garlic than a farmer in Saskatchewan could make on an acre of wheat. City farming, done on a fairly large scale, can be profitable, and when well organized using a number of backyards the annual revenue can approach that of the average 40-hectare farm in BC, without any subsidies, loans or crop insurance.

Anybody with a yard can start a garden-share business. With good soil, adequate sunlight and a green thumb you can grow your own little backyard crop; charge customers a fee of something like $50 a share and they can load up whenever they want throughout the growing season. Or you can just do the work yourself and raise enough vegetables that will go a long way to reducing your grocery bill.

Of course, this all applies primarily to homeowners fortunate enough to have an available backyard. Apartment, condo and townhouse dwellers will need to find a friend with a backyard. A more modest approach is to buy several huge tubs and grow tomatoes or peppers, spices, herbs or other food plants on the balcony. Some cities are trying to be creative by planting vegetables and spices on boulevards or establishing community gardens where a city dweller can rent a reasonable-size plot and grow anything from corn to peas to peppers and tomatoes. Community gardens, in addition to their positive environmental and economic implications can be productive as well. You see a wide variety of approaches and skills as individuals and families attempt to work together to grow food. The less adventuresome stick to the basics like beans, peas, corn and potatoes, while others grow more exotic crops.

If you live on a lake, as was the case for one couple near Powell River, BC your backyard garden presents some challenges. They built a cedar log float and then constructed 4 one-by-two raised beds filled with good soil. This enables them to grow all the lettuce, spinach, carrots, onions, and potatoes they need, plus even some strawberries.

Sales of seeds, gardening equipment and gardening books are increasing, so it seems that there is "growing" interest in home food production, in addition to growing beautiful flowers. Even in parts of Canada where winter is never far away, the moment that spring really arrives a substantial number of people rush out to plant flowers or fruits and vegetables. Raspberries and strawberries are always popular, and those with patience and faith will even plant fruit trees such as apples or plums or grapes.

VIGNETTE 46

Flip-flops

FLIP-FLOPS! EVEN THE name sounds suspicious, connoting something undesirable demonstrating poor character.

The flip-flop charge was originally used to attack politicians for advocating contradictory policies, often during elections, and frequently while denying the self-contradiction. Outside of politics the use of the term is not necessarily a pejorative. A scientist or mathematician can often obtain some experimental results or logical proofs that cause him/her to flip flop on a previously held belief. Famous flip-flops in history include the social networking giant Facebook, which has some history of backtracking. On various occasions, the company has made an important move—especially with regard to user privacy—only to reverse its course after a public outcry.

When it comes to footwear, flip-flops or thongs are an open type of outdoor footwear, consisting of a flat sole held loosely on the foot by a Y-shaped strap, like a thin thong, that passes between the first (big) and second toes and around either side of the foot. Flip-flops may also be held to the foot with a single strap over the front of the foot rather than a thong. Flip-flops are popular with those who enjoy being barefoot but need to wear shoes, because they allow the foot to be out in the open but still constitute a "shoe" for wear in places such as restaurants or on city streets, and can be quickly and easily removed. They are also popular because they are easy to carry and come in an assortment of colors and patterns.

A few years ago a women's sports team, after winning a national championship, visited the President of the United States, and some team members were criticized for wearing flip-flops at this formal and auspicious occasion. And rightly so, I would say. More recently a County Council in Ontario decided that women were no longer allowed to wear flip-flops at work. They determined that flip-flops did not give a professional impression to their clients, and that the wearing of flip-flops would indicate to one and all that the staff did not take their jobs seriously. Flip-flops may be worn at the beach or at the pool, and on those occasions when a lady has just had a pedicure, but they should not be considered as a universal piece of footwear, at least that's what I say. They may be comfortable and cute, at least to some, but attractive and professional they are not. Not that I worry much about this, but you may have gained the impression that this is a minor pet peeve of mine.

The humble flip-flop suffers from other problems. It has been said that women wear high heels for status, and the multitude of designs and designers of ladies' shoes provide many opportunities to demonstrate varying degrees of wealth, status and taste. But there isn't much surface area on a flip-flop to display status or wealth, and some would say they automatically demonstrate poor taste! In many developed countries flip-flops are typically treated as annual or seasonal, short-lasting footwear. Depending on the material makeup of the shoe, some pairs of flip-flops last a year or less. Most people don't bother to repair flip-flops because they are very inexpensive and easily replaced. These disposal habits may pose an environmental problem because most flip-flops are made with polyurethane, which cannot usually be recycled in small amounts. Because of growing environmental concerns, some companies have begun to sell flip-flops made from recycled inner tubes or car tires, as well as sustainable materials like hemp or cotton.

There is one good thing to say about flip-flops. In most developing countries, rubber flip-flops are the cheapest footwear available, often costing less than a dollar. Measures have been made to reduce cost even further so that even the poorest can afford basic footwear. Because of their low cost they are very widely used in these countries as typical footwear instead of as fashion wear. Despite their disposable design, street vendors will even repair worn sandals for a small fee.

Flip-flops

Here's a cryptic David Letterman-type list of things not to do while wearing flip-flops: 10) Be part of a wedding party; 9) Visit your grandmother; 8) Sit close to a campfire; 7) Teach a "Miss Manners" class; 6) Run a marathon; 5) Hike on a mountain trail; 4) Work in an office (see above); 3) Work at a construction site requiring steel toed boots; 2) Ride a bicycle; and finally 1) Visit the president of the U.S. (see above) or the Canadian Prime Minister.

VIGNETTE 47

Who Decided that Doctors Should Have Poor Handwriting?

IN MY EARLIER life, many many years ago, I was a pharmacist, so I have a bit of experience with this question. Sometimes, together with one or two colleagues, we would ponder a prescription to try to puzzle what the doctor had written. Some medications were fairly common, and we could determine what was being prescribed and what directions were being given. Fortunately, doctors have pre-printed prescription pads with their name on top, and on rare, embarrassing, occasions we would need to phone the doctor and quietly ask about the medication or the instructions.

I saw an article in the Vancouver Sun complaining that Canadian doctors' garbled scribbles on prescriptions and hospital charts continue to put patients at risk. But nurses also appear to have this problem. A nurse in Nova Scotia was reprimanded for illegible handwriting since his penmanship on nurses' notes and charts was unintelligible. A free course was offered to hospital staff, teaching them to write legibly at a speed suitable for a hospital's hectic pace. Of course, poor penmanship is not limited to the medical profession, but the impact of misreading notes and prescriptions here can immediately be dangerous. As much as possible, pre-printed forms and the use of computers can minimize the amount of handwriting that any professional has to do. The Canadian Institute for Health Improvement estimated in 2010 that medication

Who Decided that Doctors Should Have Poor Handwriting?

errors affect more than 1 million patients each year and that 700 patients die from medication errors.

Now there's help on the horizon. A number of hospitals, including one in BC, have introduced barcode technology that enables health care workers to prescribe medication and order tests. A prescription is entered into a computer, sent to a pharmacist who validates the prescription electronically and then places the medication on a computer equipped drug cart which is sent back to the patient's unit where nurses scan a unique barcode to confirm the drug, the dose and correct time against a barcode on the patient's wrist. Sounds like a good system to me and it might even save money as well as lives!

The word "prescription" comes from the Latin word "praescriptus". It has the prefix "pre" which means "before" and "script" which means "writing" so a prescription has to be written before a drug is dispensed or compounded. Historically, prescriptions were written in Latin and many are still written that way today using various Latin abbreviations. There are several reasons for this. Latin is more concise than other languages, so along with the various abbreviations doctors have standard shorthand that they can use. Secondly, it makes prescriptions able to be written and filled worldwide, since physicians all over the world know the Latin names and instructions. Doctors and other professionals in a wide variety of fields including the arts used to speak Latin, but obviously this is no longer the case.

Rx is the abbreviation for "recipe", which is the Latin word for "take". That's how doctors used to communicate prescriptions to pharmacists/apothecaries, when a prescription had to be made up. "Compounded" we used to call it, as in preparing an ointment or a solution with several ingredients, rather than just counting out tablets or capsules. The "take" is an instruction to the pharmacist, as in, "Take some willow bark and boil it up, and have the patient drink it." You may encounter some other Latin phrases on your prescriptions, like "qid", short for *quater in die* four times a day. Other common abbreviations are h.s. or *hora somni* (at bedtime); p.r.n. or *pro re nata* (as needed); and q.h. or *quaque hora* (hourly). Ask your friendly pharmacist some time to explain your prescription when you take it in to be filled, and you might ask him or her to explain why they charge a dispensing fee, in providing and utilizing their professional expertise.

Pharmacists do earn their dispensing fee, and not just by being able to read poor handwriting. With the assistance of their computer systems they can track all of your medications and also take the time to answer your questions.

VIGNETTE 48

Do You Know as Much About Canada as Americans Know About the U.S.?

AMERICANS ARE NOTORIOUSLY uninformed about Canada. We lived in Michigan for a few years and, for most of the people that we knew there at that time all they knew about Canada was that winter cold fronts came from north of the border. U.S. local or national weather maps back then just showed a gray, unlabeled area north of the 49th parallel presumably inhabited by some kind of hardy native people. For summer vacation, folks in Michigan would drive to Florida or even to Los Angeles which is about 3100 km or 1920 miles for them, but it would never cross their minds to drive 75 miles to Detroit and cross the border into Ontario. One medical student, knowing that I lived in Ottawa (wherever that was), said he had met someone from Toronto (population then perhaps 2 million) and he asked if I knew him.

But my impression is that Americans know more about the U.S. than we as Canadians know about Canada, even though—partly in self defense—we need to know a fair amount about the U.S. (when you sleep beside an elephant you best know a bit about it). How much do you know about Canada? Here's a pop quiz to serve as a partial indicator:

1. Who was the first Canadian-born Governor General? 2. What is the western most capital city in Canada? 3. How many Prime Ministers has Canada had? 4. Name Canada's first Prime Minister. 5. Name the only 2 Canadian politicians that have died by assassination. 6. What's the

capital of Nunavut (bonus point if you can spell it)? 7. Name Canada's only officially bilingual province. 8. What is the symbol of Canadian Football League supremacy called? 9. Name the eastern most location in Canada from which Marconi received the first trans-Atlantic message. 10. Name the two cities that have hosted "Expo". 11a. Who is Canada's Head of State? 11b. Who is the Chief Justice in Canada's Supreme Court? 12. What Canadian city hosted the meetings that led to Confederation? 13. What year did Canada become a country? 14. Who was the Opposition Leader from the west when Canada adopted the maple leaf as our flag? 15. What Canadian Prime Minister won the Nobel Peace Prize? 16. What is Canada's system of government called? 17. What is Canada's highest mountain? 18. Name Canada's first National Park. 19. Who was voted in 2004 as the Greatest Canadian? 20. Name Canada's longest river. 21. What Canadian city is directly south a of a major US border city? 22. What is the name or address of the Prime Minister's residence in Ottawa? 23. When did BC join Confederation? 24. What is Canada's "winter game" and what is its semi-official "summer game"? 25. How many seats are there in the House of Commons as of 2011? A bonus question: Which province is completely separated from the North American mainland?

How many questions did you answer correctly? Do you feel the need to do some reading and Google searches about Canada? I didn't know all of the answers either. OK, question # 3 isn't fair, but I didn't want anyone to get 100 percent. Multiply your right answers by 4 to determine your percent grade. The answers are given below, but don't peek until you answer all of the questions.

(1) Vincent Massey (2) Whitehorse (3) 22 (4) John A. Macdonald (5) D'Arcy McGee (1868) and Pierre Laporte (1970) (6) Iqaluit (7) New Brunswick (8) The Grey Cup (9) Signal Hill (St. John's), NL (10) Montreal and Vancouver (11a) Queen Elizabeth (i.e. Either the Queen or King of England) (11b) Beverly McLachlan, as of 2011 (12) Charlottetown, PEI (13) 1867 (14) John Diefenbaker (15) Lester Pearson (16) A Parliamentary Democracy or a Constitutional Monarchy (17) Mt. Logan in the Yukon (18) Banff (19) Tommy Douglas (20) Mackenzie River (21) Windsor, Ontario, south of Detroit (22) 24 Sussex Drive (23) 1871 (24) Hockey, Lacrosse (25) 308. Bonus

Do You Know as Much About Canada as
Americans Know About the U.S.?

answer: Prince Edward Island (Newfoundland doesn't qualify because it is actually Newfoundland and Labrador).

We have a great country. Let's be informed and let's be proud of our freedom and our characteristics!

VIGNETTE 49

You Know You're Getting Old When ...

YOU KNOW YOU'RE getting old when a middle aged lady gets up to give you her seat on the bus. Or when the restaurant waiter automatically tells you about the Seniors' Specials on the menu. You may be old if you remember Canadian Airlines and when they served real food, like steak, on real china. Or when the little old lady that you help across the street is your wife, and your kids are starting to look middle aged. Isn't it embarrassing when you walk into a room and forget why you're there or you start talking about Pierre Berton's books and no one knows what you are talking about? You may be past your "best before date" if you think curling is an exciting game or if you look forward to an evening at home. And then get winded just playing cards. You know you're getting old when you remember seeing steam powered trains that weren't just refurbished tourist attractions, or you remember every kid walking 3 miles to school (uphill both ways).

You're really getting old if you know what a "Bennett Buggy" is, or when John Diefenbaker was Prime Minister, or if you remember the FLQ crisis. Do you draw blank stares when you wonder out loud why Canada cancelled the Avro Arrow? Do you see incredulous comments when you say that your family never locked the front door of your house when you were growing up, even when you went away? You may be getting old if you recall the Nanaimo, BC bathtub races or the parade at the Pacific National Exhibition in Vancouver. And when the airlines agent looks directly at you and announces that anyone requiring extra

You Know You're Getting Old When ...

time boarding the flight to Palm Springs via Saskatoon should come to Gate 3, you know that you're getting old but still enjoying the good life. If your doctor told you to walk 5 miles a day and you're now 50 miles from home and lost, you know you are getting old. Or when 'Happy Hour' means taking a nap.

Time and technology may be passing you by if you have no idea what "DVR-ing" means or if you think that "Twitter" is something that birds do. You may be getting old if your kids scold you for not being a Facebook friend. You may need to do some homework if you think that iPads are something that you wear to deal with personal embarrassing accidents. Do people accuse you of having heard the Big Bang? If you need a GPS to find your way home after going shopping, perhaps you're getting old (they are now making shoes with a built-in GPS to help you, or to help others find you). We all get heavier as we grow older because we store a lot more information in our head. That's my story and I'm sticking to it.

If you remember your mother using a ringer washer and you know what a washboard is, you're not a spring chicken anymore. The same goes for you if you drive up to a curb-side mailbox and order a cheese burger with fries, or if you turn your left signal light on when leaving your street and it stays on for the whole trip. You may be getting old if you remember having to get up off the couch to change the TV channel and the TV was black and white with only 3 channels. How about when you throw a wild party and the neighbors don't even realize it? How about when you're on vacation and your energy runs out before your money does?

Here's an interesting fact: 2/3 of all people in all of history who have ever reached age 65 are alive today. I guess that's why there are so many seniors around today! The average life expectancy in Canada in 1910 was only about 50 years, while today it's about 78 for men and 83 for women.

It's not only that people are living longer today; a big reason for this dramatic change is that so many more children survive childhood. One reason for this is that vaccinations have eliminated most infectious diseases, plus we have better nutrition, better hygiene and better health care. These changes mean that we now have other challenges in helping seniors cope with life as they grow older.

VIGNETTE 50

Important and Light Weight Philosophical Questions

WHY IS IT considered necessary to nail down the lid of a coffin? When being sentenced to death, by lethal injection, why do they sterilize the needle? Can a hearse carrying a corpse drive in the carpool lane? If a person with multiple personalities threatens suicide, is that considered a hostage situation? If a funeral procession is at night, do they drive with their lights off? How important does a person have to be before they are considered to have been "assassinated" instead of "just murdered"?

Have you ever imagined a world without hypothetical situations? Why is the man who invests your money called a broker? Why do banks charge a fee on "insufficient funds" when they know there is not enough money in your account? You know that indestructible black box that is used on airplanes? Why don't they make the whole plane out of that stuff? If a bus station is where a bus stops and a train station is where a train stops, why is my desk at work called a work station? Whose cruel idea was it for the word "lisp" to have an "s" in it?

Why is it that doctors call what they do practice? Why do doctors leave the room while you change? They're going to see you naked anyway. How does the guy who drives the snow plow get to work? Where do forest rangers go to get away from it all? How is it that we put man on the moon before we figured out it would be a good idea to put wheels on luggage?

Important and Light Weight Philosophical Questions

Why is it that people say they "slept like a baby" when babies wake up and cry every two hours? Why doesn't glue stick to inside of the bottle?

Why is it that when you're driving and looking for an address, you turn down the volume on the radio? If a deaf person has to go to court, is it still called a hearing? If the police arrest a mime, do they still tell him he has the right to remain silent? When sign makers go on strike do they write anything on their signs? What's another word for "thesaurus"? If a book about failure doesn't sell, is it a success? Is there another word for synonym?

Why do toasters always have a setting that burns the toast to a horrible crisp that no decent human being would eat? What do chickens think we taste like? What do people in China call their good plates? Why does a round pizza come in a square box? What disease did cured ham actually have? Are vegetarians allowed to eat animal crackers?

Why doesn't Tarzan have a beard? Why does Superman stop bullets with his chest, but duck when you throw a revolver at him? Why does Goofy stand erect while Pluto remains on all fours? They're both dogs! If Wyle E. Coyote had enough money to buy all that ACME stuff, why didn't he just buy dinner? If a cow laughed, would milk come out of its nose? After eating, do amphibians need to wait an hour before getting OUT of the water?

Why isn't there mouse-flavored cat food? If you throw a cat out of the car window, does it become kitty litter? When dog food is marketed as being new and improved and "tastes better", who did the taste tests? If a parsley farmer is sued, can they garnish his wages? Why isn't phonetic spelled the way it sounds? (I'm hukt on fonix).

Can you cry under water? Can fat people go skinny-dipping? How much deeper would the ocean be if sponges didn't live there? If you're sending someone some Styrofoam, what do you pack it in? When you open a bag of cotton balls are you supposed to throw the top one away? Why is "bra" singular and "panties" plural? Why do you have to "put your two cents in" but it's only a "penny for your thoughts"? Where's that extra penny going to?

If electricity comes from electrons, does morality come from morons? So what's the speed of dark? Ever wonder what the speed of lightning would be if it didn't zigzag? Do the "Alphabet Song" and "Twinkle,

Twinkle Little Star" have the same tune? Why did you just try singing these two songs?

Why is it that ladies over age 60, who presumably have more time, have short hair requiring minimal attention, while younger ladies raising kids and holding down jobs have fancy hairdos?

VIGNETTE 51

Living in the Far North or Other "Isolated" Places

WE HAVE IT pretty easy living in metropolitan areas like Abbotsford or Vancouver. We just get in our car or hop on a bus or call a taxi and in a few minutes or perhaps an hour we comfortably arrive at our destination. But I noticed during the 2011 election that the Canadian Health Minister represented a riding that spans three (!) time zones and a land mass the size of Western Europe. Talk about daunting geographic challenges. A "door to door" campaign would require charting a plane and perhaps a dogsled. Another complicating factor in the north during elections is generating publicity, partly because one can't buy advertising on CBC radio, which until now has been the main electronic medium for many people in these isolated areas of our country.

I saw another article recently proposing building a 1,300 km lifeline road from Gillam in northern Manitoba, past Churchill, past Rankin Inlet all the way to Baker Lake. Right now, airplanes are the only way to reach the isolated communities in Nunavut. The economy of Nunavut would benefit dramatically from this all-weather road. Some estimates are that the Nunavut economy would then gain at least $400 million in economic activity over the next few years. Another benefit is that such a road would decrease the cost of living and the area would lose its feeling of isolation. Plans also include a hydro right-of-way along the highway which would allow the various tiny hamlets along the road to decrease their dependence on expensive diesel generators. Another important issue in the far north is that of Canadian sovereignty, so having a useable road all year would increase Canadian presence in this area.

There are other federal electoral districts, like Skeena-Bulkley Valley and Prince George—Peace River in BC that are absolutely huge. Skeena-Bulkley Valley has an area of about 324,000 square km with about 92,000 people, while Prince George—Peace River is 237,000 square km (105,000 people); together they constitute about 60 percent of the entire area of BC. Imagine the different challenges faced by the Member of Parliament for each of these 2 districts. To meet their constituents, the MP would need to visit Smithers, Stewart, Houston, Burns Lake and dozens of other towns and villages (the population density is less than 0.3 people per square km). Compare this to Abbotsford with 359 sq. km and almost 150,000 people (and three MPs) or Vancouver Centre with about 124,000 people squeezed into 16 sq. km. Canada's democratic system is more or less based on equal representation with 1 vote per person, which in part explains the huge area of these northern electoral districts and you can see that the federal government has worked pretty diligently to make this representation more or less equal. However, it's clear that these huge districts present special challenges for the MPs in meeting their constituents. Representation by population is a key to democracy, but access to MPs is also important and in isolated areas this is more difficult because of the huge distances and severe weather conditions. Social media such as the Internet using e-mail, Twitter and Facebook are starting to have a positive impact, even in the far north.

Of course even Skeena-Bulkley Valley and Prince George—Peace River pale in comparison to the task faced by the MP from Nunavut mentioned above. Nunavut is the largest parliamentary riding in the world—it has 1,935,200 sq. km but only about 34,000 people and is home to the northernmost permanently habitable place in the world (Alert). I don't think you can "hop in your car" and visit many of your constituents in Nunavut! In addition to the geography - distance factor, weather plays a huge role as well. Nunavut has land borders with the Northwest Territories on several islands as well as a tiny land border with Newfoundland and Labrador. It also shares maritime borders with the provinces of Quebec, Ontario, and Manitoba and with Greenland. Suppose you want to go to the capital city, Iqaluit. Well there are a couple

Living in the Far North or Other "Isolated" Places

of airlines, namely "Canadian North" and "First Air" that fly there from Ottawa on a semi-regular basis at a pretty substantial cost if the weather cooperates. If you want to go to the metropolis of Inuvik (population 3,000) you can fly there from Edmonton when the weather is nice. The Nunavut license plate is known worldwide for its unique polar bear shape. It was originally created for the Northwest Territories in the 1970s but adopted by Nunavut in 1999 when it became a separate territory, giving us some idea about the ruggedness of life in that vast land.

VIGNETTE 52

Social Media and Computers Are Changing the Way We Live

AN AUGUST 2010 report indicated that about 13 percent of Canadians use Twitter; this amounts to about 4.5 million people. Tweets are limited to 140 characters, and while much of the communication is trivial and even nonsensical, Twitter can be used to inspire discussion as well as provide entertainment value. But even if you don't have Twitter you can use the Internet and services such as Google to see what's happening in the tweeting world. TV and radio have been using sound bites for years and it could be said that Twitter is just another format summarizing events. Facebook reports that approximately 17 million Canadians have established profiles on its social networking service, which is a phenomenal number.

Politicians are now widely using social media to advance their platforms. Successful political campaigns in the future will likely integrate traditional media and these digital formats. Social networking is now part of what many of us do, and impacts on how we live and interact with each other. And, of course, it has changed the way many people shop, and advertisers know this. Many folks use Craig's List or Amazon.com or EBay, and Google and YouTube, among others, make huge profits by selling advertising.

And there are new hazards. Besides the curse of the "pop-up window" many individuals who are integrity challenged also have learned new ways to scam people. "Worms" and "viruses" have acquired new meanings thanks to hackers and others like WikiLeaks with motives ranging from profit, to challenging authority, to who knows what. We depend on financial institutions, retailers (and their employees) and other companies to guard our financial and personal information, but sometimes this data finds its way into the wrong hands, either deliberately or by lack of vigilance. A different hazard involves employees using an inordinate amount of time on Twitter at work to send personal messages or share corporate information, or they may surf the Internet or dialogue on Facebook, in effect stealing time from their employer. A major corporation fired an employee who used their company's account to criticize the corporation and their industry. Another type of problem is that employees may harass or intimidate co-workers on Twitter. Unlike general Internet use, which is reasonably private unless you are a sophisticated hacker, Twitter posts apparently don't disappear. Thus, Tweets containing offensive or incriminating information, pictures and links may become evidence in legal cases or in support of corporate actions.

When was the last time that you consulted an encyclopedia? The majority of people today will go to Google and Wikipedia, although this presents some element of risk since almost anyone can submit "data" to Wikepedia, and there is no guarantee of accuracy. Interestingly enough, many individuals today will also (or even first) consult their friends on Facebook about a situation or problem rather than checking out an expert or professional. Perusing blogs is a way of checking out opinions. If you want facts the onus is on you to assess the sources and separate fantasy, bias and prejudice from information that can be verified and documented. Yahoo Answers, Facebook Questions and Quora are other sites where you can find information on important issues as well as things you really didn't need to know. Plenty of expertise, both valid and questionable, is actually on display, with topics ranging from where to eat after the hockey game, to how to grow the best tomatoes, to how to build a nuclear particle accelerator.

Libraries and books are changing as well. You can buy e-books or you can check them out from your local library onto your computer for something like two weeks. Many newspapers and magazines are also available online, and some have gone out of business because of a reduction in hard copy circulation, but the switch to reading newspapers online has not yet become a stampede. We have seen a marked reduction in investigative journalism in the past few years as newspapers, and TV in particular, become cheering venues for a particular bias. They apparently often report undocumented sources rather than digging and investigating the real issues. Part of this may be because of our insistence on instant information and instant gratification in society today.

VIGNETTE 53

Big Brother Is Watching

WE ARE FILMED every day by carefully placed cameras as well as opportunistic cell phone cameras—at the bank, in many stores, at gas stations, on various streets, sometimes by the police and often by our fellow citizens. Politicians can be brought down, entertainers can be embarrassed, criminals may be caught in illegal acts, and unfaithful spouses sometimes exposed by activities captured by cell phones or video cameras. Ordinary folks voluntarily and often with great abandon share their private lives with "friends", sometimes hundreds of "friends" on Facebook or YouTube. Just to help things along on a somewhat more sinister level, we have WikiLeaks. It has released literally millions of documents, some exposing trivial undiplomatic comments by diplomats sending messages not intended for public exposure (what were they thinking?), as well as documents with varying degrees of national security. In an effort to make air travel safer we encounter huge inconveniences at our airports as big brother scans not only our luggage, but also our bodies.

Privacy used to be a concept and a right that we believed to be of great importance. Governments now have "privacy watchdogs" to prevent or address the most obvious abuses, but often it's difficult to protect people from themselves. A related issue of huge importance is the safety of kids who may unwittingly make themselves vulnerable by sharing their deepest secrets and activities and pictures on social media with friends, but also with the rest of the world, including sexual predators.

"Big Brother" in a free country like Canada or the U.S. is one thing, and something to be concerned about, but it's of even greater concern in countries where personal freedom is either an illusion or something that can be capriciously snatched away. Now we have to worry about something called "botnets"—networks of infected computers that can be controlled by a master computer. Such compromised computers are infected with a malicious code or malware, which is sent to computers through various means like email attachments, spam emails, video downloads, and music downloads.

George Orwell wrote the provocative and prescient novel "1984" in which he dramatically described and predicted the activities of "big brother" watching ordinary citizens in sinister and secretive fashion. "Big brother" used deliberate euphemisms as a masquerade for their activities (as I recall, the "Department of Peace" was in charge of waging war, for example). Experts mocked many of Orwell's predictions at the time, but his forecasts were generally uncannily accurate. I suppose this just goes to show that law-abiding citizens need to be vigilant to ensure that authorities don't over react in their efforts to protect them.

You might ask, what's next? Technology seems to be moving along faster than our ability to control it or at least recognize all of the implications of these changes. There already are devices which can scan your handbag, briefcase or laptop as you walk past a sensor. Our written messages and our phone calls can be easily monitored and recorded. Now what about our thoughts??!! Will some enterprising physicist or physician or engineer be able to invent a device that records our thoughts as we walk past their detectors or when they surreptitiously bug our house and telephone? It sounds preposterous, but is it within the realm of possibility in the near future? It seems that the answer is yes. Researchers at the Centre for Brain and Mind at the University of Western Ontario employed functional Magnetic Resonance Imaging to successfully predict the action of participants' hands *before* they'd moved a muscle, and other scientists have been able to determine what nouns (things) that a person was thinking about. Your mother probably told you to be careful what you say but a wise mother now also needs to say, "Be careful what you think!"

VIGNETTE 54

Pure Science vs. Applied Science vs. Junk Science

IN SIMPLE TERMS, there are two types of research—"pure research" or curiosity research that tackles the basics of how things work but without any specific result in mind, and "applied research" which more or less directly supports industry by developing new products and techniques. Pure research may seem esoteric and an expensive luxury, but it's the basis of many discoveries and is crucial in maintaining a strong scientific base and keeping world class scientists working in Canada. Pure research is therefore needed so that applied research can take these ideas further and develop products that support industry and society in general.

One of the great Canadian institutions, unknown to the great majority of Canadians, is the National Research Council (NRC), largely based in Ottawa. Scientists working for NRC have a reputation for world class research that results in both new products and new technology. NRC has been very strong in both pure research and applied research, and some famous scientists like Nobel Prize winner chemist John Polanyi and physicist Gerhard Herzberg have worked at NRC. Some scientists believe that pure research should be almost exclusively done in universities, and that organizations such as NRC should stick to research that supports and enhances strategic industries. While it has done excellent curiosity research, NRC has functioned as a bridge between universities and industry, and it has demonstrated that it can and should do both types of research.

An example of NRC success in applied research was the work they did in helping build the billion dollar canola industry in Canada. There are other nationally important projects that also require applied research techniques; these include increasing the productivity of wheat farming, developing techniques that enable algae to soak up carbon dioxide from emitters, developing bio-composite materials, and developing new techniques to build electronic circuits.

Junk science, sometimes called pop science, occurs when scientific facts are distorted, when risk is exaggerated or discounted, or when science is adapted and warped by policies and ideologies to serve another agenda. Just as real science has prestigious awards for those scientists who make incredible break-throughs and contributions to our knowledge base (such as the Nobel Prizes for Chemistry and for Physics and the Albert Lasker Award for Medicine), someone has proposed the "Rubber Ducky" award for those who contribute to obfuscation and sophistry while confusing science with advocacy. The 2011 winner of a Rubber Ducky was New York City Mayor Michael Bloomberg who, perhaps with good but misguided intentions, used his influence to ram through various ill-advised health related policies. Rubber Ducky winners typically include activists, politicians, journalists, non-government organizations (NGOs), media outlets and quacks. You can see that junk scientists are often not qualified scientists at all, but real scientists are human too, with their own biases, and they sometimes (frequently?) get caught up in reporting only data that supports their theories and preconceptions. Scientists working for tobacco companies, for example, are often purveyors of junk science.

Two United Nations "investigators" were nominated for the Rubber Ducky award for using two methods that junk "scientists" frequently practice. These officials misquoted someone and also took biased information from a special interest group or think tank, and then offered this data as scientifically established fact. This approach has also been used in the climate change debate, for example, where an advocate of global warming used information from Greenpeace plans as real data that supported their advocacy views. Junk scientists may often strongly criticize anyone who doesn't accept their conclusions and reports. An insidious example of junk science at work was the full

Pure Science vs. Applied Science vs. Junk Science

panel notice that I recently saw on a gallon carton of milk in the U.S. It indicated that this national grocery chain would not sell any milk or milk products where the milk came from cows that had been treated with BGH (bovine growth hormone, which increases the milk production of cows). It didn't seem to matter that they were required to also carry a notice basically indicating that the FDA had determined that milk from such cows was safe (in part because BGH rapidly breaks down when consumed by humans so that there are no known effects in humans) or that it's becoming increasingly difficult to maintain milk production in the U.S. (or Canada) due to commercial and residential encroachment of farm acreage. But the pressure by idealistic activists essentially forces dairy producers to avoid the use of BGH. I understand that Canada is in a similar position. Health Canada has determined that there is no logical or scientific reason why BGH could not be used, but they have decided that the public isn't ready to accept its use.

Another problem with junk science is that public pressure and politics often overwhelm real science. Activists often are not fully or correctly informed, and they often knowingly or unwittingly spread misleading or false information. One example is the informal campaign against BC Hydro's "smart meters" where claims of hazardous radioactive emissions and bizarre claims of "big brother" were common. Another example was the environmental pressures to prematurely ban incandescent light bulbs on the basis of incomplete evidence. The unstated mantra of junk scientists may be said to be "my mind is made up; don't confuse me with the facts." A powerful force behind much of this junk science is conspiracy theory, where individuals will believe their fears and mistrust authority rather than scientific facts.

VIGNETTE 55

Unions Aren't What They Used to Be

AND THAT'S PROBABLY a good thing. Unions tell us that they don't like strikes, but you would never know it by the vehemence and commitment frequently seen on the picket line. Sometimes unions have public sympathy while on strike, but probably more often they do not. This is particularly the case when public sector unions strike, because the public perceives, generally with good reason, that these workers already have good wages and excellent benefits. The challenge faced by unions and union leaders today is probably tougher in some ways than it has ever been. The recession of 2008 and the slow economic recovery since then have resulted in a public attitude of restraint and a further erosion of support for striking unions.

A further complication comes from events south of the border. States like Wisconsin and Minnesota, facing bankruptcy, have taken dramatic and drastic actions to reduce or strip collective bargaining rights of public sector unions. Their cutbacks even included, dare we say it, school teachers. And the public, already nervous about both their own financial condition as well as that of local, state and federal governments, has generally been supportive (excepting of course, political opponents and union militants). Some analysts think that public sector unions are still in a strong position, but if the U.S. and other countries don't address their serious debt problems these unions could be under further attack, and this tough approach could also spill over into Canada. In the U.S., the average salary including benefits for private sector workers in 2011 was about $60,000, but for federal employees the average was more than $120,000. This disparity seems unacceptable at any time, and especially so in difficult economic times. I don't know about the Canadian figures, but it's safe to say that the average salary for public sector employees is substantially higher than for the private sector.

Globalization and the modern world have also resulted in new challenges for unions. New technologies and international competition have reduced the clout unions once had. Unions that once had the opportunity to significantly impact a given industry could essentially force the employer to agree to many of their demands at the bargaining table. We regularly saw this in the auto industry as the unions forced "sweetheart packages" on employers. But today a strike against Ford or GM basically creates opportunities for Toyota or Audi or Volvo or a dozen other suppliers. Union leaders are beginning to recognize this, and they generally are more conciliatory and more cooperative than they were 20 or 30 years ago.

One thing I don't understand is why more companies and more unions don't take the Canadian company West Jet approach more often in making workers part "owners" and therefore part of the solution rather than part of the problem. The other thing union leaders could do, if they really want to be cooperative rather than confrontational, is to focus on working with management to become more productive and more dependable. Investing in people, rather than focusing on demands and benefits would seem to be a good thing for more unions to consider. I understand that two of the factors limiting more than a few Canadian companies (the Vancouver Port Authority may be one example) is that productivity is low compared to some other countries. They are considered to be undependable due to the almost omnipresent threat of strikes. Making concessions is an anathema to unions, but giving up some of the sweetheart issues that they wrung out of short-sighted managers when times were good would both increase public support and improve the job security of workers and managers. Requiring three people instead of one person to perform three different but related tasks, such as happens in the construction industry, for example. Workers sometimes getting paid for not working, but still demanding to be there at the job, or receiving double or triple overtime under dubious circumstances, are other examples.

Unions still sometimes do decide to take the militant aggressive, "burn all the bridges" approach. I spoke with the manager of a grocery store that was part of a national chain, and he said that the union at his store has been on strike for 30 months! The "good" or most conscientious employees have all gone to jobs in other grocery stores or related jobs, but the militants are still walking the picket line. Probably the main reason they are doing so is that the national union is paying them full salary (which is tax free because they already paid taxes on their salary and union dues), so life may be boring but then they don't need to work at all. Maybe they are happy enough pointlessly walking around outside for a few hours each day. The company has recognized that the union is so intransigent that it's not possible to have productive talks with them. The national union has apparently decided that things have gone on long enough, but the "local" won't give up and evidently has not even passed on the last several company offers to the strikers for a vote. One might think that somebody should give their head a shake.

VIGNETTE 56

Snail Mail Isn't Dead Yet, But....

DO THE UNION members at Canada Post have a death wish? Or some of the senior managers, when it comes to that? In spite of competition from a wide range of sources, the union workers insist on going on strike every few years rather than working proactively with management to carve out new business models. They need to recognize that the conventional dominant status of Canada Post in delivering ordinary mail is being threatened. Most of the mail we receive at our house really could be provided by other means. The biggest volume of mail consists of bills and junk mail, and most utilities and other companies are working quite hard to have all their billing and payments done on-line which saves money for both the customer and the company. Much of banking today is done electronically. More than 90 percent of the advertising junk mail we receive goes directly into the garbage or recycling. Surely there is a better way and no, I'm not asking for more TV commercials.

While the government was concerned about the economic impact of the Canada Post strike in June 2011, this strike did not have the same paralyzing effect that a postal strike did 30 years ago or even 10 years ago. The Internet and e-mail, plus the increasing use of couriers, have created significant competition for the postal workers but they seem not to recognize this. It will be interesting to see what happens in the next 10 years. Will postal workers and Canada Post work together, or will Canada

Post disappear? There's good reason to believe that they can survive, but this will be difficult unless the workers in particular recognize that they need to adjust to work in the 21st century. During the 2011 strike I heard of an apiarist who had ordered something like $20,000 worth of Queen Bees that were critical for the coming year's operation. The shipment of bees was held up in some postal distribution center and unless delivered in 48 hours they would all die. Canada Post was the only service willing to deliver this type of cargo at the present time, but you can bet your indexed pension that this businessman will have found an alternative shipping mode by the next time postal workers go on strike. Once a service is lost, it's highly unlikely that it will be regained. One of the weaknesses of a postal union strike is that they can't withdraw all services, since they are basically forced to handle welfare cheques and pension cheques to low-income Canadians and senior citizens at the risk of totally losing public support.

Canada Post still does a lot of business, so there is a place for optimism if management and unions cooperate. I saw a figure indicating that Canada Post delivers over one million letters and parcels each day, which works out to about 30 billion per year. This seems like something to build on. A 59-cent stamp in 2011 enables a person to send a letter across the city or across the country and have reasonably high assurance that it will reach its destination in a few days. Couriers are great, but they are pretty expensive for small businesses. Plus Canada Post currently has a monopoly on first-class mail, and if they lose this by demonstrating that they are undependable, the workers best start thinking of transitioning to new careers. Various levels of government also use Canada Post, perhaps because of cost factors, whether it be Census Forms or tax notices, or election information, but even governments could seek other options if postal workers are obdurate and ordinary mail is undependable.

VIGNETTE 57

The Organic Movement— Is It Just Another Form of Activism?

NOT THAT ACTIVISM is a bad thing, but I find that activists often are emotionally attached to their cause and have difficulty acknowledging, or a reluctance to check out, facts or differing opinions in an effort to promote and protect their cause.

Selling organic food is big business, perhaps as much as $30 billion in Canada and the U.S. It may be said to have a noble objective, that is, to grow and sell wholesome food, but it goes off the rails on several counts. For starters, the organic food industry states, or at least implies, that non-organic food is laced with pesticides or other chemicals. Organic food is claimed to be healthier, tastier and more wholesome than conventional foods, but the evidence for this is generally subjective and scientifically unsubstantiated. All food is "chemical," and at least in North America there is evidence that the vast majority of foods do not contain pesticides or other unwanted or unapproved chemicals.

The Organic Food industry in Canada, and BC in particular, is represented by The Certified Organic Associations of BC (COABC), which is an umbrella association for certification bodies providing certification accreditation and leadership in the development of organic food production throughout British Columbia and Canada. COABC works to ensure program credibility, facilitate domestic and international trade, and to promote the overall growth of the organic food community in BC. COABC is committed to promoting the importance of organic agriculture, which helps ensure the long-term health and vitality of our region. COABC seeks to educate farmers transitioning to organic production; advocate for organic consumption; provide current research and vital information for consumers, producers, stakeholders and the media; and interface with federal and provincial governments.

COABC has set up a rigorous certification system for organic growers, but there still are problems in demonstrating that any given food package labeled as "organic" is in fact organic. Random surprise testing is an essential tool that government food regulators or industry advocates could use, but it isn't clear that this is being done by those certifying food products as organic. Another challenging problem is that much of our organic food is imported from countries such as Mexico and China, and as such is not as likely to conform to Canadian growing requirements to meet the organic designation. Testing is expensive, but not unreasonably so, and a chemical test that can screen for at least 200 pesticides can cost in the range of perhaps $200. The premium that consumers pay for organic foods should cover this.

Another problem that detracts from the credibility of the organic "movement" is that much of the impetus or *raison d'etre* is emotional and non-scientific. One specific indication of this is that many organic activists refuse to use modern seed varieties because they assume that old varieties are better since they supposedly possess natural pest resistance. My understanding is that there is no scientific evidence that old varieties have better resistance just because they were bred before chemical fertilizers and pesticides were used.

VIGNETTE 58

Why Is It So Hard to Export or Implement Democracy?

WINSTON CHURCHILL, PROBABLY the greatest statesman of the 20th century, famously is quoted as saying, "It has been said that democracy is the worst form of government except for all the others that have been tried." We find this easy to believe in Canada and the United States, but many countries around the world either don't understand the concept or have such vested power systems that democracy isn't really on the radar.

The U.S. and Canada have sacrificed many lives in countries like Iraq and Afghanistan, while making precious little progress in helping install a democratic government. The plan or theory was that by protecting the local population and providing some stability, a secure environment would be established where services could be provided and the population would embrace the government, and eventually democracy. But success has been elusive in both of these countries. President Karzai's power in Afghanistan depends on his maintaining the loyalty of his associates—ministers, governors and local warlords and district leaders. Guess how he does this? Well, it seems mostly with money and guns. His regime is said to be rife with corruption and the governmental system, such as it is, generally isn't competent or strong enough or interested enough in delivering the various services needed to give ordinary people confidence in honest democratic government. Karzai, in the spirit of self-preservation, can hardly be expected to move towards a democratic system by reforming the system and losing power in the process. The United States gave him huge, almost autocratic powers that limited any opposition parties when they worked with him to draft a Constitution, so it will be interesting to see what happens when his term is completed.

Iraq is in somewhat better shape but is still quite unstable. The 2003 invasion by the U.S., which led to the removal of Saddam Hussein's dictatorship, resulted in a transitional government and is now described as a representative democratic republic. The business of "nation building" is quite tenuous. It's not clear what will happen after the approximately 45,000 U.S. soldiers as of May 2011 left—it's quite possible that Iraq will not be able to cope and the "democratic" government will fall to another autocratic regime. One report has it that the U.S. has lost more than 4,400 lives in trying to help Iraq, and spent more than $900 Billion, with more than $9 Billion worth of equipment and supplies reported as lost or stolen.

Then there's Egypt. Many westerners cheered when President Mubarak was pushed out of power. The virtual dictatorship of Mubarak was replaced by an authoritarian Armed Forces Supreme Council, but democracy still seems distant since the army was reluctant to relinquish control. Political activists who dared to criticize the government have been jailed rather than have their allegations investigated. Many Egyptians desperately want change that includes civil liberties and freedom of expression, but as of April 2011 the Supreme Council didn't rush to enact reforms and was slow to initiate a corruption prosecution against Mr. Mubarak, although they did eventually bring charges against him. Ordinary citizens won't have a voice in Egypt's political process until the military gives up its absolute hold on power, and at this time there is little indication that this will happen quickly. There was considerable fear in 2011 that democratic elections would result in a government that would initially be democratic but would slide into being radical Islamic rule that would soon impose severe Sharia law. Tunisia did have free elections in 2011 and apparently achieved the rare case of a moderate Islamic government; time will tell if this is able to endure.

It appears that it is almost impossible to implement democracy in any of these countries if there is no religious freedom. When governments insist on implementing restrictions that penalize minorities or use coercion and violence or cripple religious reform and freedom, the outcome usually is extremism and authoritarianism rather than democracy. It appears that, even apart from humanitarian considerations, religious liberty is necessary for the stability and longevity of democracy and for the defeat of religion-based terrorism. Even a well-known democracy like India is regularly rocked by religious extremists that threaten its stability. Syria is another country where the regime is in trouble but there is no evidence to suggest that democracy is on the horizon. Saudi Arabia seems even further away given the government's stranglehold on the country.

Russia may be a more complicated example. When the USSR collapsed in 1991 following the changes led by Mikhail Gorbachev, there was some hope that democracy would follow. But the events that followed, led by Vladimir Putin and others, illustrated the difficulty of implementing true or full-fledged democracy. Putin was "elected" as President from 2000 to 2008 before switching roles and becoming Prime Minister, but he allowed wide spread corruption to continue as his friends and other entrepreneurs became hugely wealthy while the lot of the common people improved very little. The elections of 2011 were widely considered to be rigged as Putin manipulated the election to hold on to power, and it appeared to many observers that Russia was sliding again into an authoritarian government. The lure of power seems just too tempting, and Putin was apparently contemplating again becoming President and staying in power at the Kremlin for another 12 years.

One could list many other countries where, theoretically at least, democracy and a free exchange of ideas is desirable but real progress achieving this is tenuous at best and realistically very problematic. More than a few African countries, and some South American nations, espouse democracy, but the magnetic draw of power often means that those in power desperately work at staying in power.

VIGNETTE 59

Why Do We Have Fan(atic)s?

I HAVE NEVER understood the compulsion that some people have about collecting autographs. Or the intense adoration that many have for the rich and famous like rock stars or movie stars or athletes, or even, dare I say it, political leaders. Or the determination that apparently happy people have to always watch "ET Tonight." Part of this phenomenon, I suppose, is fine, where individuals are simply interested in people who live a different kind of life that is exciting, probably wealthy and perceived to be happy. But part of it also seems to be a felt need to live vicariously in the reflected glory of some other person who is beautiful or successful, and to be part of something larger than ourselves. We see this particularly in the adulation of movie stars when ladies buy magazines at the check-out counter. We see it in professional sports when guys call in to sports talk shows to talk about "our team" and are excited when they win and devastated when they lose.

A fanatic is marked by excessive enthusiasm for and intense devotion to a cause or idea. It may be expressed as a belief or behavior involving uncritical zeal, perhaps for a religious or political cause, or sports, or with an obsessive enthusiasm for a pastime or hobby. One philosopher defined fanaticism as "redoubling your effort when you have forgotten your aim." Winston Churchill said that a fanatic is one "who can't change his mind and won't change the subject." In any case the fanatic displays little tolerance for contrary ideas or opinions. There is a difference between a fan and a fanatic, although sometimes the difference is pretty murky. The enthusiastic behavior of a fan towards a given sports team or movie star or subject is different from the behavior of a fanatic who exceeds or violates prevailing acceptable or "normal" social behavior. The behavior of a fan may be odd or eccentric, but it is normally just considered as devotion to a team or a person or a cause.

Then there are people (fanatics) who are considered to be eccentrics or cranks. These are individuals who hold positions or opinions that are determined by society at large to be ludicrous and/or probably wrong, such as a belief in a flat earth. In contrast, the subject of the fanatic's obsession may be "normal", such as an interest in sports or religion or politics, except that the scale of the person's involvement, devotion, or obsession with the activity or cause is abnormal or disproportionate and may lead to stalking or harassment.

There are many expressions of fanaticism. Sports fanaticism is one of the most common as we often see high levels of intensity surrounding sporting events. This needs to be qualified because in many cases the person uses sports activities as a testosterone fuelled, or alcohol fuelled masculine excuse for brawls. English soccer fans used to have this reputation, and we have also seen it periodically after dramatic Vancouver Canuck losses or victories.

Political or ideological fanaticism can lead to ethnic or racial supremacist fanaticism, which is a terrible thing to see, as in what has happened in various parts of the world including some eastern European countries, Africa, and North America. Leisure fanaticism (perhaps exhibited by an excessive drive to jog or participate in other exercise programs) is generally much less dangerous, but may be expressed by excessive levels of intense enthusiasm and zeal shown for a particular leisure activity. Consumer fanaticism is what we see when "groupies" will do almost anything to relate somehow to a particular person or group or even an emerging fashion. Religious fanaticism may be expressed as an extreme form of fundamentalism.

In some cases, fanaticism is a subjective evaluation. What constitutes fanaticism in someone else's behavior or belief is determined by the core assumptions of the one doing the evaluation. Politicians running for elected office sometimes come close to desperation and even fanaticism when they make sweeping statements and generalizations, twisting what their opponents are actually saying.

So, be a fan! Enjoy cheering your sports team, enjoy the love of the game, enjoy seeing the underdog overcome obstacles, and enjoy witnessing greatness. Enjoy watching your favorite movie stars, or enjoy supporting your church and beliefs, and enjoy participating in your favorite leisure activities, but don't cross the wide grey line to fanaticism.

VIGNETTE 60

Grandchildren Are More Fun than Children!

YOU PROBABLY HAVE heard the saying, "If we had known that grandchildren were so much fun we would have had them first." Being a parent is a wonderful feeling and has many rewards, but with young kids and teenagers there are always subtle or direct pressures and challenges. Are we doing the right thing, are we disciplining properly, are we providing the best opportunities, what if we can't afford to send them to the orthodontist, whatever will we do when they become teenagers; how do we protect them from illegal drugs; how do we teach them to make the right choices; well the list goes on and on. But the role of a grandparent is very different—spoil the grandchildren for a few days when they come to visit, and then heave a sigh of relief and combined relaxation with a feeling of nostalgia when you see the tail lights of their family car disappear down the street. Even if the kids live close by and you can see them and the grandchildren regularly, you can always send them home after each visit. And all of these experiences can be very rewarding.

Within weeks of getting married, my sweetheart and I moved from Edmonton to Ottawa. We treated this as a huge adventure as we started our lives together, which it was, and I had little thought of what my parents and my new wife's Dad were thinking as we drove away. Eventually three kids came (one at a time) and I started to think as a parent. For several years our summer vacations consisted of driving back to Alberta (three days and two nights each way), and we saw the joy that grandparents have as they see their kids, and it seemed, especially their grandchildren. Eventually we moved to the Vancouver area and together with our kids we could see our families more often, and when each of the kids spent one year in Bible College in Edmonton they developed an even closer relationship with their grandparents.

Now we have eight grandchildren of our own and we are especially blessed since all three of our kids live within an hour of our home. We see them all pretty much every week as we help with babysitting or picking kids up after school, or just hanging out. It's a fabulous opportunity to see beautiful young lives again as they grow and experience life through family, church, sports, music and school. Last year all 16 of us had a great experience as we spent a long weekend together camping, eating, swimming, eating, playing sports and just hanging out, with everyone getting along. And we all spent a beautiful Sunday afternoon in the fall as a photographer took thousands of family pictures in myriad family groupings. My favorite picture is that of the 16 of us holding hands in a wide line as we walked towards the photographer. A few years ago we went to Disneyland, but there were only 14 of us at that time. Experiences like this are life enriching.

VIGNETTE 61

The Space Shuttle and Our Need for Galactic Adventures

WHERE WERE YOU on July 20, 1969? That was the day that Neil Armstrong uttered those famous words: "One small step for man, one giant step for mankind," as he became the first human to walk on the surface of the moon. The Russians, with Yuri Gagarin, were the first to catapult a man into space, but the Americans, perhaps feeling that national pride was at stake, quickly expanded their space program and with the enthusiastic support of President Kennedy determined to be the first on the moon. It must have been pride or hubris, perhaps combined with a genuine search for knowledge, because going to the moon didn't really have any profit implications.

Years from now you probably will not remember the date of July 8, 2011, which was the date of the last American space shuttle mission. The space shuttle was developed in 1981 and shuttles were launched 135 times in those 30 years, carrying over 800 men and women into space.

The astronauts receive all of the attention and adulation, but behind them is a large army of support personnel. In 2005 there were more than 17,000 people working for the shuttle program. They included welders, simulation supervisors (who put the astronauts through every imaginable problem and stress inducer in training prior to lift-off), fuel-tank builders, crawler engineers (the giant crawlers carry the orbiter, fuel tanks and booster rockets to the launch pad), spacesuit technicians, rocket-booster builders (including the explosive charge that ignites the rocket just before liftoff), and the inevitable math and computer geeks and engineers. But shortly after the last shuttle flight in July 2011 there were only about 1,000 workers. The shuttles were space haulers of satellites and equipment, they were used to repair or retrieve disabled satellites, they served as a rescue vehicle, and after the mission the shuttle could return back to earth. The shuttle program taught scientists how to deal with stratospheric lightning bursts, how to design productive "spacewalks" and perhaps even more importantly it taught those running the space program how to successfully deal with a myriad of unexpected problems. It's sort of like watching someone do something for you as opposed to doing it yourself; then when you need to do it alone you are generally not prepared for all the variables and problems that invariably happen. If the Americans stop "doing space exploration" for themselves, how much capability and expertise will they lose?

The redundancy of the shuttle is partly because of the development and completion of the International Space Station. As a joint project between the United States, Russia, Japan, Canada and Europe, the International Space Station is said to be the largest and most complex international scientific project in history. It orbits the earth at an altitude of 250 miles; this allows the station to be reached by the launch vehicles of all the international partners and to provide capability for the delivery of crews and supplies. The station measures 356 feet across and 290 feet long, with almost an acre of solar panels to provide electrical power to six state-of-the-art laboratories. Canada is providing a 55-foot-long robotic arm to be used for assembly and maintenance tasks on the Space Station.

Astronauts spend a variety of times living on the station, usually between four to six months. They carry out a range of experiments, observations, maintenance, and construction on the station. The unmanned Russian Progress spacecrafts regularly visit with supplies, as do manned Soyuz spacecraft. Since the U.S. manned space shuttle will no longer deliver astronauts, supplies, and parts for the construction project, the Russians will provide this service for a while.

A concern by many in the U.S. is that the shuttle is being phased out without a replacement at hand. Given the complexities of the science and engineering expertise that is required, but which will be lost or dissipated by the NASA cutbacks, it will be difficult and costly to reassemble the

The Space Shuttle and Our Need for Galactic Adventures

critical mass of knowledge when the U.S. once again decides to expand their space program. It is recognized that the International Space Station will play a key role in space research for the next few years and do some of the things that the shuttle did. It is also understood that the shuttle is being phased out now since its full range of capabilities is no longer needed. The monies that were spent on the shuttle need to be diverted to pay for developing the next generation of spaceships, as yet not well defined (although there is a distinct possibility that this money will be used instead for debt financing and other pressing U.S. needs). This makes sense except that one can't help be concerned that the U.S. has lost its vision for space exploration that it had 50 years ago. For the next few years at least, the U.S. will be depending on hitching a ride with the Russians (at more than $50 million a seat) or with private industry back and forth to the space station. Several countries, including China and India have plans to send people to the moon, so the U.S. reputation as a space leader will suffer.

Can the U.S. or mankind in general, afford not to explore? Can we just build fences around ourselves and just worry about our families, jobs and debts and pensions? History and human nature suggest that we must always be asking questions.

VIGNETTE 62

Some of My Favourite Quotations

I SEEM TO be a collector of quotations. Perhaps this is because I have so few original thoughts of my own, or because I often read a book or a newspaper with a pen at hand. But I often admire the wisdom and humor of others, so here's a short list of quotes from some famous people and some not-so-famous folks.

Right is right, even when no one is doing it—G.K. Chesterton.

God is the God of the second chance. And the third chance … and the hundredth chance—Nicky Gumbel.

The conventional definition of management is getting work done through people. Real management is developing people through work - Agha Hassan Abedi.

The evening of life is as beautiful as the morning, and far richer, for it has fulfilled itself—Taylor Caldwell.

A person who has had great parents is twice blessed, first when they are alive, and second for the memory of them—Taylor Caldwell.

Humility is not thinking less of yourself; it is thinking of yourself less—Rick Warren.

Prayer is not overcoming God's reluctance. It is laying hold of His willingness—Richard Trench.

Home computers are being called upon to perform many new functions, including the consumption of homework formerly eaten by the dog—Doug Larson.

Some of My Favourite Quotations

Everybody knows how to raise children; except the folks that have them—P.J. O'Rourke.

It's too bad that all the people who really know how to run the country are busy driving taxi or cutting hair—George Burns.

Spring is when you feel like whistling even with a shoe full of slush—Doug Larson.

Live every day like it's your last; 'cause one day you're gonna be right'—Ray Charles.

The only thing harder than living alone is living with another person—I. Trobisch.

Man must have just enough faith in himself to have adventure, and just enough doubt of himself to enjoy them—G.K. Chesterton.

A man's work is from sun to sun, but a woman's work is never done—Anon.

Sometimes I wake up grumpy, but usually my husband gets up first—Melody Zylla.

Love is the triumph of imagination over intelligence—H. Mencken.

Love is composed of a single soul inhabiting two bodies—Aristotle.

True love comes quietly, without banners or flashing lights. If you hear bells, get your ears checked—Erich Segal.

Love doesn't make the world go round. But love is what makes the ride worthwhile—Franklin Jones.

Three groups of people spend other people's money: children, thieves and politicians. All three need supervision—Dick Amey.

Complaining is good for you as long as you're not complaining to the person you're complaining about—Lynn Johnston.

It's tough to make predictions, especially about the future—Yogi Berra. The future ain't what it used to be—Yogi Berra.

You can observe a lot by just watching—Yogi Berra.

Grief can take care of itself, but to get the full value of joy you must have someone to share it with—Mark Twain.

You must do the things you think you cannot do—Eleanor Roosevelt.

Whoever said you can't buy happiness never took their kids to the candy store—Anon.

Whoever said you can't buy happiness forgot about puppies—Gene Hill.

The older I get, the better I used to be—Lee Trevino, professional golfer.

If I ever need a brain transplant, I want one from a sportswriter because I'll know it's never been used—Dave Ritchie, former BC Lions Coach.

I don't make jokes. I just watch the government and report the facts—Will Rogers.

And one final word: "That's All Folks"!!!—Bugs Bunny.

VIGNETTE 63

Sports—No More Dynasties, But More Greed

BEING PART OF what is euphemistically called the Middle Class, I don't understand the thinking of many of our sports heroes. Let's take the 2009/10 NHL Chicago Blackhawks as an example, but my thoughts could apply to almost any winning team. In June 2010 this hockey team won the Stanley Cup with its great balance of high scoring forwards, gritty 3rd and 4th liners, excellent defensemen and a promising young goalie. The team kept their top stars the next year but lost several of their best tenacious forwards and defensemen. The 2010/11 team was pretty much a shadow of the previous year, just barely making the playoffs on the last day of the regular season and losing in the first round of the playoffs. A couple of their players earned in the $1.6M range in 2009/10 and moved (or were moved) to another team that had lots of 'salary cap room' for perhaps $0.5M or $1M or $2M more. Here are my questions: How much greater is your take-home pay when you earn $2.3M per year as opposed to $1.6M, or $6.3M vs. $4.5M? Would you rather play on a team that had an excellent chance to win another Stanley Cup for $1.6M or play for a team in the southern U.S. for $3M that has no chance of even making the playoffs and can't fill the arena and your take home pay isn't much higher? How much is enough? Is one's lifestyle significantly enhanced if you earn $2.3M instead of $1.6M?

It's not that I'm against million dollar salaries for professional athletes; after all they are being paid what the market will spend. What I don't like is the lack of loyalty exhibited by modern day athletes. In baseball they call them "rental players" and there are examples of players playing for 3 different teams in 3 different years. "Free Agency" has clearly been a great thing for athletes. After a certain number of years they can sell themselves to the highest bidder and move to a new team. On the surface this is fair. As their lawyers and agents would say, anyone can do this so why would we discriminate against athletes? I suppose one answer is that it's a bit different for a fellow or lady earning $52,000 per year, since they are unlikely to be able to double or triple their salary by moving across the country at that earning level, with all of the associated moving costs.

The implementation of a salary cap in hockey came about after the last strike/lockout in 2004/05 in an effort to limit salary costs and increase parity between teams. While parity has happened, we have also seen an even greater movement of players related to free agency. After winning a championship, players want a raise and in some cases deservedly so, but the end result is that teams are often pretty much dismantled the next year. It seems that we are unlikely to see the dynasties of the Montreal Canadiens or New York Islanders or Edmonton Oilers any time soon. Maybe this is OK, but I think the average fan spending more than $100 per game for a ticket, parking and a hotdog could expect some team consistency as well as some reward for general managers and coaches who build up a team.

No National Hockey League team has repeated as Stanley Cup champions since 1997/98, and it doesn't seem like there will be a repeat champion anytime soon. A very few dynasties still do happen in some sports. Perhaps the best (worst?) example is Soccer's English Premier League where Manchester United (one of the richest teams in the world) has won the championship 11 times in the past 20 years. There are 20 clubs in the league, but only the same three or four teams each year have any realistic chance of winning the title. There is no salary cap, the justification being that a cap and parity would hurt the top clubs' competitiveness in playing in Europe. Four of the top clubs play in the "Champions League" each year, in addition to playing in the Premier League, and they make a lot of extra money this way. In contrast, the three bottom feeders are relegated to the minor leagues while three top lower-level champions move up to the premier league, generally only to be relegated to the minors a few years later since they don't have the cash to pay top level players. One other league where dynasties are at least possible is major league baseball in the U.S. Baseball doesn't have a salary cap, but it has a "luxury tax" where a club pays a tax if they exceed a given figure. But this doesn't really work since the really rich clubs (owners can be "greedy" just as much as players) have more money than the U.S. Treasury. They pay the tax and spend as much as they want. In 2011, for example, the New York Yankees' payroll was about $203M, while the next highest club is $172M and the lowest of the 30 teams was only $36M. It hardly seems like a level playing field. I guess the fact that the big spenders don't win every year is a tribute to the competitive nature of the players. The Yankees have won the "World Series" 27 times out of 40 appearances, more than any other club.

Professional basketball players are the highest paid athletes in the world, with an average salary in 2008 of $5.4M, and this figure doesn't include the very lucrative endorsement deals, bonuses, or other financial perks that come from being a NBA player. And yet they don't seem satisfied! The "salary cap" does seem to work in establishing at least some semblance of parity. But one interesting note I read was that roughly 60 percent of NBA players go broke within five years of retirement. I don't know what the figures are in other sports, but I suspect money management is an issue everywhere. Bestowing instant millionaire status on impressionable young men desperate to prove their worth comes with some hazards, even with money manager "agents" taking a big cut.

One argument for high salaries in professional sports is that careers are quite short, and certainly this is true. But when players earn $50,000 or more per game you might conclude that the system is out of whack. Why is it that we pay professional athletes more than doctors or more than teachers? The short answer is that's what the market will pay. The long answer, including a discussion of our societal values, is more complicated. When President Hoover complained that Babe Ruth earned more one year than the President, Babe Ruth is said to have replied, "That's because I had a better year than you did."

VIGNETTE 64

We Live in a Multi-Faith Society

I READ AN article in the April 2011 issue of Christianity Today where Ed Stetzer described the four world religions that represent about 75 percent of the world population. Recent surveys show that there are 2.1 billion Christians, 1.5 billion Muslims, 900 million Hindus and 376 million Buddhists. Stetzer questioned the value of interfaith dialogue because the [false] premise of such discussion seemed to be that there are no fundamental irreconcilable differences between these religions. He went on to suggest that the basic belief in each of these religions was the idea of God, so he compared this belief in each of these four religions.

Hindus can believe there in one god, millions of gods, or no god at all. Ancient Hindu writings teach that the soul is god, meaning that god is in each person and each person is part of god. Buddhists, on the other hand, do not believe in any god since they believe that religious ideas and especially the idea of god, stem from fear. For most orthodox Buddhists the concept of god is at best unimportant and at worst an oppressive superstition.

Islam says in the Qur'an that "He is Allah, the One and only. Allah is the Eternal, Absolute. He begets not and he is not begotten. And there is none like unto him." A Muslim primer for children says that "Allah is absolute and free from defects and has no partner. He exists from eternity and shall remain eternal. All are dependent on him, but he is independent of all. He is father to none, nor has he any son." In contrast Christians believe there is one God who is creator of the world. He is a personal God, a conscious, free, and righteous being. And he is not only a personal God but a God of providence who is involved in the day-to-day affairs of creation. He is a righteous God who expects ethical behavior from each of us. He expects his followers to live out their beliefs by loving him with all their heart, soul, mind, and strength, and by loving their neighbors as themselves. God, while one essence, also reveals himself in three persons: Father, Son and Holy Spirit.

Stetzer's conclusion was that we if can't agree on the basic definition of God or his character, we can't say that all the major religions are on the same path toward the truth about God. We have to acknowledge that we live in a multi-faith society since these religions have radically different concepts of the future, eternity and the path to get there. Recognizing this also means allowing adherents of other faiths to live out their beliefs and convictions without conflict. Respecting and understanding someone's beliefs does not mean accepting those beliefs. In a multi-faith world we recognize that we are not worshiping the same God or gods, and we should not be offended by our mutual desire to proselytize one another so long as we do so respectfully. Stetzer suggested that mutually exclusive religions can co-exist if we let each religion speak for itself (and stop generalizing the excesses of other religions), talk with and about individuals, not generic faiths, respect the sincerely held beliefs of people or other religions, and grant each person the freedom to make his or her own faith decision. I suppose it would be good if this thinking would also apply to the "new atheists" who vehemently advance their own belief system or faith positions (though they fiercely resist calling them that) while being critical of Christianity.

Intellectuals and scientists have long been scornful of religion or belief systems. Karl Marx said that religion was the opiate of the people, meaning I suppose that he thought it helped reduce the pain of everyday life and lulled people into a sense of well-being. A hundred years ago the German philosopher Fredrich Nietzsche, and other great thinkers, predicted the death of God and a decline of religious belief, but recent surveys suggest that religious belief is on the rise today. Although they may not realize it, scientists and philosophers probably demonstrate more faith in doing their research than the average person, since they make assumptions and accept ideas that they can't always prove. There are two options to the question of what holds the world together—God or matter. Both positions involve faith. Cultural anthropologists, for example, say that there is no one religious tradition that is right (this statement alone is self-contradictory since they imply that their statement or tradition is correct) and that bringing Christian viewpoints to others is hateful. Faith is a universal assumption; when someone says that your belief system is wrong, they are making a faith statement that their own belief system is correct.

VIGNETTE 65

What Are You Afraid of?

A PHOBIA IS an irrational, persistent fear of certain situations, objects, activities, or persons. The main symptom of this disorder is the excessive, unreasonable desire to avoid the feared subject. When the fear is beyond one's control, or if the fear is interfering with daily life, then a diagnosis under one of the anxiety disorders can be made. It is generally accepted that phobias arise from a combination of external events (i.e. traumatic events) and internal predispositions.

The internet tells us that the most common fear is said to be the fear of insects, especially spiders (arachnophobia). As with all phobias, the strength of the associations means the individual must not actively pursue the consequences of encountering spiders, and outsiders should not in any way tease or play with the phobia. Snakes kind of fit in here too, or at least should generate a healthy respect. The fear of being evaluated negatively in social situations and the fear of public speaking are also very common. Some people get faint or even pass out if they have to give a speech. A little bit of nervousness is a good thing, but some folks really do need help here, or must find a job and interests that don't require public speaking. I've always admired those people who can speak without any apparent nervousness, and often without extensive notes.

Thanatophobia, or fear of death, is a relatively complicated phobia, and many people are afraid of dying. Some people fear being dead, while others are afraid of the actual act of dying. However, if the fear is so prevalent as to affect your daily life, then you might have a full-blown phobia. Many people's fear of death is tied into their religious beliefs; Christians who believe in eternal life often fare better in facing death. Another factor is the death of family and friends and not wanting to think of the world without them. Necrophobia is the fear of dead things. Agoraphobia involves intense fear and avoidance of any place or situation where escape might be difficult, or help unavailable, in the event of developing sudden panic-like symptoms. This is closely related to claustrophobia, which is the fear of being trapped in small confined spaces.

Monophobia is the fear of being alone. Mono is Greek for one, single or alone. People can experience loneliness for many reasons and many life events are associated with it. The lack of friendship relations during childhood and adolescence, or the physical absence of meaningful people around a person are a few causes for loneliness. The loss of a significant person in one's life will typically initiate a grief response; in this situation, one might feel lonely, even while in the company of others.

There aren't as many white knuckle fliers as in years gone by, but the fear of flying (aerophobia) may be a distinct phobia in itself, or it may be an indirect combination of one or more other phobias, such as claustrophobia or acrophobia (a fear of heights). It may have other causes as well, such as agoraphobia (fear of open spaces). It is a symptom rather than a disease, and different causes may bring it about in different individuals. Perhaps the fear of heights is due to the thought of falling that far and hitting the cold, hard ground and breaking bones. Some suggest that humans are land based creatures and therefore being in high places gives many people a bit of a scare. Brontophobia, the fear of thunderstorms, is also on the top ten list of phobias. Perhaps the fear of lightning should also rank right up here. Other common phobias include the fear of the dark, horror movies, fear of falling in love, fear of swimming (or water in general), fear of needles, and fear of failure.

It may seem a bit odd, but some people have a fear of success. This can happen in every walk of life when you feel that you are not worthy of your success or position. It almost invariably comes in the way of self-esteem and sense of responsibility. The best way of getting rid of such a thought is to feel that you are a normal human being, not superior or inferior to others. Many folks have a fear of financial problems. Tallulah Bankhead said she had been rich and she had been poor, and rich was better.

There are some bizarre phobias. Coulrophobia is an abnormal or exaggerated fear of clowns. Sufferers sometimes acquire a fear of clowns after having a bad experience with one personally, or seeing a sinister portrayal of one in the media. The weird appearance of the clowns, swollen red noses and unnatural hair colors, makes these persons look so mysterious and treacherous. Adults who are victims of coulrophobia know what they fear is completely irrational and illogical, but they can't escape the circumstance. Gymnophobia is a fear or anxiety about being seen naked, and/or about seeing others naked. Gymnophobes may experience their fear of being nude in front of anyone, or only certain people, and may regard their fear as irrational. This phobia often arises from a feeling of inadequacy that their bodies are physically inferior, particularly due to comparison with idealized images portrayed in the media.

The fear of Friday the 13th is called paraskavedekatriaphobia, a word that is derived from the concatenation of the Greek words meaning Friday, thirteen, and *phobia,* and is a specialized form of triskaidekaphobia, a fear of the number thirteen. Panphobia, also called omniphobia, is a medical condition known as a non-specific fear; the sufferer finds themselves in a state of fear but with no known target, and therefore no easy remedy. It has been described as "a vague and persistent dread of some unknown evil." This fear is often seen as a secondary condition to schizophrenia. I'm not making this up, but the fear of cotton balls is called sidonglobophobia. Luposlipaphobia is my favorite phobia. This is the fear of being pursued by timber wolves around a kitchen table while wearing socks on a newly-waxed floor. This is actually a fictional phobia which was created by Gary Larson, the author of the Far Side comics.

I am afraid that I have ordinateurmechanio-reparaphobia which is a new French term for being afraid to repair things; my wide ranging inability and fear includes the fixing of computers, bicycles, household appliances and most anything else you can name.

VIGNETTE 66

Prayer—Why and How Do We Pray?

WHY DO WE pray? Well, one reason is that Jesus prayed; another reason is to develop and grow a relationship with God. We pray because God wants us to tell him what is on our minds. Even though he sees and understands our situations better than we do, he still wants to hear from us in our own words. We pray to God the Father in Jesus' name in the power of the Holy Spirit.

Does God always answer prayer? The short answer is yes. The more complicated answer is that sometimes the response to our request is yes, sometimes it is no, because our requests are often not good in themselves, or not good for us or for others, directly or indirectly, immediately or ultimately. Sometimes the answer is "wait." There are some barriers that may delay or inhibit positive answers to our prayers. Unconfessed sin in our lives, disobedience, or a refusal to forgive someone may act as barriers. Having wrong motives or misunderstanding the will of God may also cloud our expectations.

Our obsession as to whether prayer works is the wrong question. We know prayer works. The real question is—are we prepared for God's answer? Why put so much effort into praying if we already know that God is generous and that He already knows our needs? If we ask this question, maybe that's our way of asking what is the minimum required of me to get my prayers answered. We pray not only because God answers prayer, but also that we might recognize and receive His answer, know how to respond, and feel His presence.

Prayer—Why and How Do We Pray?

How should we pray? There are lots of ways! The Lord's Prayer is a good approach—not rattling through it in 20 seconds, but pausing to ponder on each phrase in a prayerful attitude. "Our Father in heaven, Holy is Your name; Your kingdom come; Your will be done on earth as it is in heaven; Give us this day our daily bread; Forgive us our sins as we forgive those who sin against us; Lead us not into temptation but deliver us from evil; For thine is the kingdom, the power and the glory. Amen."

Another way to pray is to use the "ACTS" acrostic. A is adoration as we praise God for His nature and His goodness to us. C is confession as we ask for forgiveness of our sins. T is thanksgiving as we give thanks for our many blessings. S is supplication where we ask for things we need and as we make requests for others. Many of us start with supplication and often don't make it to adoration or thanks. Others "throw up emergency prayers" to God, asking for help when they get into a serious jam.

When should we pray? The Bible says pray without ceasing, which doesn't mean that we should drive or walk around with our eyes closed, but rather that we are always in an attitude of prayer with a recognition of God's constant presence with us. It's also essential to often pray alone as we commune with God. Having a regular time to pray is a good idea, and we should select the best time of the day for us (some are morning people, others may prefer other times). Martin Luther said something to the effect that he prayed two hours per day if he was busy and three hours per day if he was really busy. Dr. Luther was a saint and I've never managed to pray an extra hour when I was busy, never mind really busy. Finding a good place to pray is also a good idea, and some find it helpful to have a pen and paper at hand to quickly write down distracting thoughts so they can be dealt with later.

There are at least four styles of prayer and the one we use may depend on our circumstances and our personality. The first style is *devotional*, or scripture-based prayer. Many people like to read a Bible verse or passage that opens a way for them to talk to God. They often begin their prayers by reading a Psalm. The second style is *spontaneous* prayer. This prayer comes from our hearts almost without thinking, and it is usually a prayer of thanksgiving or supplication: "Thank you, God, for the beautiful sunset;" or "Please help Aunt Mary today during her surgery." The third style is *conversational* prayer. Many people have long and involved conversations with God, where they talk things over with him, always sure that he is listening. This style also works well in groups.

The fourth style is *intercessory* prayer, where we ask God to care for another person or part of his creation. Some people have suggested there is another type of prayer, namely the prayer of *action*. Some people pray best while walking or moving or doing something. Sitting still is hard for them, and they need to add movement to their prayers. For them doing something (an act of kindness, for example) can be praying.

VIGNETTE 67

Millions, Billions and Parts Per Quadrillion

MILLIONS, BILLIONS AND Parts Per Quadrillion. It sort of sounds like a nursery rhyme, but in economics, government, chemistry and various other walks of life these terms have very important implications. We sort of know what a million is, but after that we lose sight of how big these numbers are. A billion is a thousand millions. I don't know because I haven't ever seen it, but I've been told that a million dollars is a stack of brand new $1000 bills nine inches high. A billion dollars is a stack of brand new $1000 bills 750 feet high, or the height of a 75 story office tower. Rather a startling description of how much bigger a billion is than a million!

In 1944, Canada's national debt was about $8 billion, in 1975, $19 billion, in 1981, $91 billion, and in 1993, $458 billion. Canada's debt in 2011 was about $560 billion; this costs us something like $40B a year in interest payments. To whom do we owe the money? Mostly to other Canadians, who hold Government of Canada savings bonds, Treasury bills or other similar financial vehicles. About 25 percent of the debt is held by foreigners. Is the debt coming down? It did by several billion dollars a year for a while, but it rose again after the 2008 economic turndown. The federal government in 2011 ran a deficit budget, which means the national debt is still increasing. It will be several more years before they have a balanced budget, much less start paying down the debt.

Here are some American examples of how easily we throw a billion dollars around. In early 2011, the U.S. debt was growing by $3 billion per day! A Louisiana Senator asked the U.S. Congress for 250 BILLION dollars to rebuild New Orleans ... interesting number ... what does it mean? Well, there were about 484,600 residents of New Orleans (every man, woman, and child) so *each* one should have received $516,528 to rebuild their lives, homes and businesses. Or if you had one of the 188,251 homes in New Orleans, your home should get $1,329,787. But I suspect it didn't work out that way even if the funding was approved.

Imagine, now, $700 billion spent for bailing out banks in the U.S. That's enough to fund complete medical care for every man, woman and child currently alive in the U.S. for 11 years!! 50 billion to bail out the auto industry??? Washington, D.C. and Ottawa—Hello!!! Are all your calculators broken? How can you keep spending money this way? Well, one obvious answer is that we, the ordinary citizens, demand huge, extensive services from governments and we complain loudly and vociferously if these services are cut back or, heaven forbid, are stopped. One example is that, even though enrollment is decreasing in many BC school districts, school boards and especially school teachers are vehemently criticizing the "small" increases (!!) in education funding. On the federal side it seems that the Harper government has decided that doling out stimulus money is not the best thing to do, and they have taken a better approach by planning spending cuts.

Let's look at these millions and billions from a different perspective. In a food analytical laboratory they can detect chemicals such as pesticides at the part per million level or the part per billion level. How much (how little) is 1 ppm? Well, think of 1 can of soup in a case that contains 1,000 cans—that's 1 part in a thousand. Suppose that there are 1,000 cases in a boxcar. That's 1 part per million or 1 ppm. Now suppose there are 1,000 railway cars on that train so that 1 can of soup would be 1 in a billion (1 ppb). The chemist's job is to find that 1 ppm or 1 ppb of chemical. For some particularly toxic chemicals a good laboratory can detect 1 part in a quadrillion (a thousand million millions!). One ppq would be a decimal point followed by 14 zeros and then a 1 (0.000,000,000,000,001).

Millions, Billions and Parts Per Quadrillion

Another use of these "millions based" terms involves megabytes, gigabytes and terabytes in the computer industry to describe disk space, or data storage space, and system memory. A few years ago hard drive space was described using the term megabytes, but the tech world quickly moved to the gigabyte, which today is the most common term being used to describe the size of a hard drive, although terabyte is also now a common term. A megabyte is 1,000,000 bytes in decimal notation (actually the figure is 1,024 x 1,024 it's 1,048,576 bytes, but that's more detail than we need here). When you buy an 80 gigabyte hard drive you will get a total of about 80,000,000,000 bytes of available storage. Even larger terms are now being discussed; 1,024 gigabytes is 1 terabyte, then by a factor each time we have the petabytes and the Exabyte. There are more "byte sizes" coming, but you get the picture. Another indication of how the world is changing is that you can now buy a Swiss Army knife, that in addition to the scissors, nail file, screwdriver and other gadgets, now also has a flash drive that holds one terabyte of data. Back in 1993 the entire sum of Internet traffic only amounted to 100 terabytes. This one terabyte on the handy little knife is enough storage to hold 220 million pages of text, or two years of non-stop music, or at least 330,000 pictures!

A new standard is the Exabyte which is a billion gigabytes. According to the Washington Post, the global capacity to store digital information on CDs, computer hard drives, smart phones etc. was 276 exabytes in 2007. This is equivalent to having a stack of CDs that would extend from your desk to 80,000 miles past the moon! And about 94 percent of all information storage capacity was digital. The other 6 percent was in books, magazines, videotape and other analogue formats. Computers are here to stay, and our storage capacity is continually increasing.

VIGNETTE 68

Cross Country Odyssey 2006

IN MAY 2006 my sweetheart and I started our big trip across Canada at 4:30 am in our motorhome with a little red car in tow. Our first stop was at the Tim Horton's just down the street for coffee and timbits (yes, they are open that early). The motorhome chugged up the Coquihalla highway into the BC interior and we stopped at Kamloops for breakfast and $104 worth of gas. We were determined not to worry about the price of gas so it won't be mentioned again. We went through Golden and camped at Radium Valley. After breakfast the next day we went through the Tim Horton's at Windermere for coffee and then to Fernie and through the beautiful Crowsnest Pass slumbering gracefully in past mining memories, to Pincher Creek Alberta (famous for its windmills used for generating electricity), and finally we stayed at Lethbridge overnight. The next day saw us at Swift Current, and we experienced the vastness (flatness!) of Saskatchewan. Moose Jaw was worth a stop to see the underground tunnels built in the late 1880s and 90s by steam engineers to connect downtown buildings, but the tunnels were expanded for and used by Chinese laborers brought to Canada to help build the CPR. After an overnight stay in Regina we drove to Winnipeg.

After a day of rest we left about 7 am for Thunder Bay where we saw a giant statue of Terry Fox looking down the highway; this was close to the place where he had to abandon his marathon of hope across Canada because of his cancer. The whole day was spent driving around Lake

Superior; sometimes the view was quite spectacular. The Canadian Shield with rocks and trees and rivers and picturesque lakes seems to go on forever, but eventually we arrived in Sault Ste. Marie featuring zillions of mosquitoes and the St. Mary's River, which separates Ontario from Michigan. We had decided to take a short-cut through East Lansing Michigan (where I went to university) and crossed the border again to Woodstock, Ontario and Stratford-on-Avon, ending up a few days later at Niagara Falls. The city is unimpressive but the Falls are always spectacular and the small upscale and picturesque city of Niagara-on-the Lake is pretty cool. Peterborough was the next stop; the only exciting part was driving past Toronto on highway 401 with its millions of trailer trucks. Peterborough is a typical Ontario city with the ever present brick houses and a "lift-lock" canal, which is now used only for pleasure craft.

Our plan was to travel fairly quickly through the west until we got to Ottawa, after which we would take more time. We visited Hull, Quebec where we had coffee on the patio of the Museum of Civilization and a fabulous view of Ottawa including the back of the Parliament buildings, the Supreme Court and the Chateau Laurier. We toured the new War Museum (war is old, but this museum is new), the Bank of Canada Currency Museum, and then walked to the Confederation Building on the west side of Parliament Hill. We had arranged for a behind the scenes tour of parliament with our MP, which included walking on to the floor of the House of Commons and the Senate. Dow's Lake, the Rideau Canal and the National Art Museum were other Ottawa attractions that we saw, plus the Bytown Market which is like Granville Island Market in Vancouver.

The next stop was Montreal where we took a three hour bus tour, but I wouldn't recommend this even though we saw the Olympic Stadium ("the Big Owe"), St. Joseph Oratory and Mt. Royal. We visited some friends just south of Montreal where they lived on a beautiful little lake, and then we drove through the Eastern Townships to Quebec City. Original settlers were given a narrow strip of land stretching back from the St. Lawrence, and we could still see some evidence of that including the prosperous farms. Again we went on a bus tour, but we had a better time walking through the old city, and saw the impressive Chateau Frontenac and the Plains of Abraham, the many narrow

European-type streets and even the sweet smell of horses when a caleche went by. Riviere du Loup was next and we enjoyed the Parc des Chutes, and some beautiful waterfalls, plus a great view of the town. We were a bit too early in the season to see the whales, although this is a popular activity around Riviere du Loup.

We drove along the St. John River once we entered New Brunswick; the highway was excellent and the scenery was great. We set up at a very nice campground and then drove into Fredericton. I got in another good bike ride, from our campground into Fredericton along the St. John River. We had made arrangements to park at our friend's house in Moncton where they operate the Magnetic Hill B and B. He drove us to Magnetic Hill, which is an optical illusion where you press on the gas pedal to drive down the hill, and when you turn around you roll up the hill. I tried this on my bike later, and it was an odd feeling when I had to pedal quite hard "down" the hill and rolled "up" the hill with no effort.

Our friend drove around to the coast to Shediac (with the "world's largest lobster") and the spectacular Hopewell Rocks along the Bay of Fundy. The tides here are as high as 45 feet or more, and they rise very rapidly due to the narrowness of the Bay. Another highlight was a fabulous lobster feed at their house with all the fixings including corn on the cob, salad, and dessert. New Brunswick was very enjoyable and we were impressed with the big, green well-manicured lawns and the beautiful houses and occasional farms. A lot of NB is covered with trees used to make pulp and paper, but they have lots of potato farms in the western part, and the coast line is generally very beautiful.

We crossed the 15 km Confederation Bridge linking NB and Prince Edward Island and could immediately see the red earth of PEI. We camped at Cavendish, an hour away on the north coast along the Gulf of St. Lawrence, and then we drove to Avonlea Village and "Green Gables" of *Anne of Green Gables* fame. I enjoyed seeing the many potato farms and dairy farms. PEI is fascinating; almost every time we came around a corner we could see the ocean, sometimes with red cliffs and sometimes with beaches. We dipped our feet in the ocean and enjoyed scallops for dinner. There apparently are three main industries in PEI, namely farming, working on the ocean (either fishing or tours) and tourism. The impressive things about PEI are the well maintained (and usually large) houses, and the meticulously manicured lots and farm yards that make every place look like it could be in a magazine or on a postcard. Alongside many of the roads there are pink and purple lupines growing, which just adds to the beauty of the countryside. One day we took the "Blue Heron Drive" along the north central coast and drove into one area on a narrow red clay trail to the ocean where there were a couple of houses and a few cottages. It was kind of desolate and beautiful at the same time. Then we drove to Summerside and part of the western side of the island, which is a bit more rugged and with more trees, but we also saw some huge potato farms. Our next stop was a campground in Charlottetown. We toured Founders Hall, telling us about the Charlottetown Conference of 1864 involving "Canada" (the colonies of Upper and Lower Canada) and the Maritime colonies meetings leading up to the BNA Act of 1867. I also saw old Government House and the actual room where the conference meetings were held. Later we drove to a picturesque little fishing town where we talked with a fisherman who was baiting his lobster traps. At other times of the year he fishes for herring and scallops.

We left PEI on a ferry at Wood Islands for a 75-minute ride in dense fog to Antigonish, Nova Scotia. The next day we headed for Baddeck and toured the Alexander Graham Bell museum. We again enjoyed mussels, lobster, coleslaw, potato salad and dessert. The Cabot Trail has been described as the most beautiful drive in Canada, and it's almost as nice as BC's Sea to Sky highway. There were lots of pull-over viewpoints to stop, especially in the national park. Then it was on to Halifax where the first order of business was going to Peggy's Cove, which was fantastic. We saw the classic Peggy's Cove picture in real life in the sunshine, plus the lighthouse on the rocks. Mahone Bay was pretty neat too, as was Lunenburg, even though the iconic Bluenose ship pictured on our dime wasn't in port. Downtown Halifax was old and new and interesting, and there's good access to the waterfront. We toured The Citadel; the old fort from the 1860s that dominates Halifax, and saw Pier 21 where many immigrants first arrived when they came to Canada in the 1900's. We drove in the rain to Digby, and the next day we caught the ferry to St. John, NB. They had an impressive Farmers Market and we saw the "Reversing Falls," so named because at high tide the ocean tide is higher than the St. John River and actually reverses the flow of the river for a couple of hours—and kilometers—with lots of churning of the water as these forces collide. It was very interesting! Our last stop in Canada was St. Andrew-by-the-Sea in New Brunswick, a very picturesque seaside fishing town, and then we headed to Bangor, Maine.

This travelogue is about Canada (PEI was my favorite, but everywhere in Canada there is beauty), but suffice it to say that as we drove back on the U.S. interstate highways we also enjoyed seeing parts of Maine, Massachusetts, Connecticut, New York, Pennsylvania, Ohio, Indiana, Illinois (including a hectic drive through Chicago), Iowa, South Dakota and Mount Rushmore, Wyoming, Montana, Idaho, and finally Washington. We arrived home just in time to see a newborn grandson, a highlight of a tremendous trip!

VIGNETTE 69

The Resurrection of Jesus

NICKY GUMBEL OF the ALPHA Course (essentially a "Christianity 101" course) says that the resurrection of Jesus is the most documented fact in history. Christ's teachings, his works, his character, his fulfillment of Old Testament prophecy, and his conquest of death all provide evidence of his being the Son of God. I believe that Jeff Bucknam, Lead pastor at Northview Community Church, would agree since there is massive and incontrovertible evidence that Christ's resurrection was real. In Matthew chapter 28 the two Marys came to the tomb and experienced a "great earthquake as an angel of the Lord came and rolled back the stone ... his appearance was like lightning, and his raiment as white as snow. And for fear of him the guards trembled and became like dead men." The Gospel goes on to say the angel told these women not to be afraid, showed them where Jesus had lain, and told them that Jesus had risen. As they joyfully left to tell the disciples, they met Jesus, and took hold of his feet and worshipped him.

As Christians we are good at talking about Christianity as a positive experience, which is great, but we also can have confidence to talk about it as verifiable fact. It doesn't matter as much that Christianity makes us feel good as it is that Christianity is true. The attempts to deny the resurrection started right on Easter Sunday. The Chief Priests and Pharisees in Jerusalem had hired Roman soldiers to guard the tomb so that no one could steal Christ's body and say that he had come back from the dead. But their plans went awry when Jesus rose from the dead.

According to Roman law, the Roman soldiers should have been killed since they were derelict in their duty when the stone was rolled away, but the Jewish leaders paid them big money to protect the soldiers from their superior officers and say that they fell asleep during the night. Their story doesn't hold water—how did they know that "the disciples stole the body" when they (the soldiers) were sleeping? And why would these burly soldiers risk going to sleep, and even if they did sleep how was it that they wouldn't hear the massive stone being rolled away?

Several theories have been advanced to deny the resurrection and provide an alternate explanation of what happened. One is the Swoon Theory, which suggested that Jesus didn't really die on the cross, so that he just woke up in the tomb and escaped; this ignores the facts that he was beaten to within an inch of his life before being hung on the cross, then experienced the horribleness of crucifixion, and then was declared dead. Then there is the Conspiracy Theory saying that Jesus' body was stolen by the disciples. But this doesn't explain why the disciples were willing to be tortured and killed if they were just defending a lie. The Hallucination theory suggests that his followers so much wanted Jesus' promises to come back from the dead that they just imagined that this happened. The Jewish leaders could have refuted this theory just by producing the dead body, but of course could not do so. Finally, the Myth Theory says that all of the stories about Jesus' life, death and coming back to life just grew and grew until people accepted the stories as true. But the Bible gives us specific, verifiable facts, ranging from the name of the temple policeman whose ear was cut off when Judas betrayed Jesus, to the Roman centurion who saw Jesus die and exclaimed that Jesus truly was the Son of God, to documenting that the two Marys touched Jesus' feet when they first worshipped him that resurrection morning. The only explanation that makes sense is that the resurrection happened—everything that Jesus had predicted came true. Then there is the evidence that Jesus appeared to the disciples and later on at least 10 occasions to more than 500 others; making all of that up would have been impossible.

What does all of this mean? (The first thing that the disciples did when they saw Jesus the first Easter Sunday was to worship him. Perhaps we should do this as well!). The Gospels tell us what happens when we die—the Resurrection has made it possible for us to live eternally. Jesus has given victory over death.

VIGNETTE 70

Minor Political Parties in Canada

WE KNOW QUITE a lot about the major political parties in Canada—the Conservatives and the Liberals. Even the NDP is considered as a major party even though they never have achieved power federally and can make all kinds of promises showing how they would spend our money. The Bloc Québécois (BQ) is a federal political party in Canada devoted to the protection of Quebec's interests in Canada and promoting Quebec sovereignty, but its future is in some doubt. They were established in 1991 as an informal coalition of Progressive Conservative and Liberal Members of Parliament from Quebec, but are hardly a "national" party.

Depending on how you count them, there are more than a dozen "minor" political parties whose platforms range from the bizarre to mischievous to the unrealistic to the idealistic to the irrelevant. Under provisions of the Canada Elections Act of 2004, political parties are only required to nominate one candidate in order to be registered as a political party and as such candidates can be listed on the ballot alongside the party's name rather than being designated as independents. Some designate themselves as populist, and most of their leaders are strong personalities who don't play well with others, while some seem to use elections just as a platform to advance their ideas, fully recognizing that they have no hope of electing a Member of Parliament.

The Christian Heritage Party of Canada, founded in 1987, is a social conservative party that advocates that Canada be governed according to Biblical principles. The Progressive Canadian Party, founded in 2004, includes "Red Tories" who did not want to move from the Progressive Conservative Party that morphed into the Conservative Party in 2004. The Canadian Action Party (founded in 1997) promotes Canadian nationalism, monetary reform, and electoral reform and opposes "neoliberal" globalization and free trade agreements.

Interestingly there are two communist parties (the Communist Party of Canada and the Communist Party of Canada (Marxist-Leninist)); I guess this is the far left and the really far left (why can't these guys get along?). The Libertarian Party of Canada subscribes to the tenets of the libertarian movement as a hybrid of left-wing social policy and right-wing economic ideas; its eight candidates won 1,949 votes in 2004. Their laissez-faire policies include reducing government's role in the economy (like eliminating the CRTC which controls broadcasting, the Bank of Canada, and the federal income tax and sales tax), plus support for civil liberties such as free association and free speech, a non-interventionist foreign policy, and repealing the Canada Health Act.

The Marijuana Party of Canada has a pretty basic platform: end the prohibition of cannabis. Then there is the Rhinoceros Party which was founded just to satirize Canadian federal politics; a basic party tenet was to take lightly its own hare-brained electoral promises. While they fielded a surprising number of candidates in past elections they might not do so any more since the candidate fee to run in an election has been increased to $1,000. The Natural Law Party of Canada is the Canadian branch of the international Natural Law Party, the political arm of Maharishi Mahesh Yogi's Transcendental Meditation movement; its most famous member was an unknown magician.

Some minor parties no longer exist and some run just a few candidates or even no candidates in elections. Some like the Reform Party and the Canadian Alliance Party have morphed into bigger parties, in this case the Conservatives. Even the Social Credit Party had federal MPs for a few years but then disappeared from the scene when their charismatic leader left politics. Other minor parties like the Freedom Party of Canada, the Nationalist Party of Canada, the National-Socialist Party of Canada, and the feminist People's Political Power of Canada are unregistered, but still somewhat active.

The Green Party is a "major minor" political party. They advance a broad multi-issue political platform that reflects its core values of ecological wisdom, social justice, grassroots democracy and non-violence, but one wonders if they would be better off advocating their issues within one or more of the major parties. The party broke 1 percent of the popular vote in 2004 and 6.8 percent of the popular vote in the 2008 election, but only 4 percent in 2011, although they did elect one MP in a sympathetic (yuppie?) west coast riding. One of the major impacts of the BQ and the Greens seems to be to make it much more difficult for one of the major parties to achieve a working majority in parliament. I read somewhere that voting for the Greens is like voting for the Rhinoceros Party but not as much fun.

VIGNETTE 71

Signs of the Apocalypse

WHEN I TRAVEL on an airplane, which is maybe half a dozen times a year, I regularly buy a Sports Illustrated magazine at the airport to help pass the time. We used to live in the U.S., so I have maintained an interest in U.S. major college sports, and of course now living next to the U.S. we are all regularly bombarded with American professional and college sports on TV. When sitting in a micro-sized airplane east, a Sports Illustrated is easier to fold than a daily newspaper. Anyway, one of the sections in this magazine is called "Signs of the Apocalypse" describing unusual or outlandish actions or comments by athletes or sports officials.

Here are a few examples:

- Engineers in Qatar unveiled a design for a remote-controlled, solar-powered artificial cloud, which they hope to float over games at the 2022 Soccer World Cup to provide shade from the scorching summer sun.
- Two Calgary men were arrested last week after they allegedly got drunk, broke into an ice arena and stole a Zamboni, intending to use it to push their car out of the snow.
- David Ferrer, the world's No. 6-ranked tennis player, attempted to silence an infant in the stands who had cried during a serve at the Sony Ericsson Open quarterfinals by intentionally directing a forehand lob into the vicinity of the noise.

- Award-winning Brigham Young University basketball player senior Jimmer Fredette is said to have been asked by school officials to finish his degree online because his presence in classes had become too distracting.
- The New York Yankees asked the city of New York for $370 million in bonds for their new stadium the same week they signed pitcher C.C. Sabathia to a $161 million contract.
- The Cincinnati Reds baseball team was unable to hold a ceremony honoring Pete Rose's record-breaking 4,192nd hit on Sept. 11, the 25th anniversary of the play, because Rose had an appearance scheduled at a casino that day (Rose had earlier been charged with betting on his own team).
- Schools in and around Athens, Ga., were shut down the day before the Georgia-Florida university football game because so many teachers called in sick.
- In Freetown, Massachusetts, a seven year-old boy was benched at a Little League baseball game because his mother was a no-show at the concession stand (she couldn't get off work) and officials indicated that concession stand revenues were necessary to fund the league's programs and, well, rules are rules.

The term "apocalypse" generally has referred to the Bible's description of signs of the imminent destruction of the world and the salvation of the righteous. Some of the signs of the end times mentioned in the Bible include: wars on a global scale, famine, pestilence (sickness and disease), lawlessness (crime), people having no love for one another, and earthquakes. In our "me first" modern society we often lose sight of what's really important and act in foolish, selfish ways, as well as recognizing some of the personal and natural catastrophes that occur. The term apocalypse is regularly used today to describe great or total devastation or doom as in *the apocalypse of nuclear war.*

Signs of the Apocalypse events or actions are not limited to sports, as evidenced by the following examples:

- Some parents in the U.S. were selling chicken pox laden lollipops to other families who want their children to catch the virus and

contract the disease while they are young, while others held pox parties where kids could mix with those who had chicken pox.
- The senior manager of a trendy clothing company estimated that he spends 20 percent to 30 percent of his time dealing with counterfeit problems, especially scams on the Internet, since enterprising counterfeiters are selling cheap imitations that damage his reputation and his sales.
- A couple were at a marriage counselling seminar. When the speaker said that husbands should know their wife's favorite flower, the husband poked his wife and proudly whispered "I know yours. It's Robin Hood all-purpose flour."
- Death row inmates in New York and California are not allowed to smoke cigarettes because it's bad for their health.
- Illegal immigrants to the U.S., who don't vote or pay taxes, take to the streets a million strong demanding their right to break our laws. Smokers, who pay more taxes than anyone and obey the law and VOTE, are silent.
- During a town hall meeting, President Obama was asked, "Is the American Dream dead?" The president responded in part by saying that, "Absolutely not.... There is not a country in the world that would want to change places with us."
- The "I Love Lucy" show was not allowed to air on one network TV because Lucy and Ricky smoked on screen.
- Officials in Delhi, India are putting microchips (which contain a unique ID code) in snakes used by snake charmers in entertaining tourists to ascertain that the snakes have been registered by their owners.

VIGNETTE 72

Only 56 Years Ago ...

GLEANINGS FROM THE Internet yielded a few gems—comments made in the year 1955 that apparently sounded reasonable to the speakers at the time, but are now outlandishly dated. Reading these gems reminded me of the saying by the short sighted and probably pompous U.S. Patent Office Commissioner, who said in 1899: "Everything that can be invented has been invented." My memory is fading, but I can still remember 1955, so I can relate to the circumstances that led to these *bon mots*.

- "I'll tell you one thing, if things keep going the way they are, it's going to be impossible to buy a week's groceries for $10.00."
- "Have you seen the new cars coming out next year? It won't be long before $1,000.00 will only buy a used one."
- "If cigarettes keep going up in price, I'm going to quit; 20 cents a pack is ridiculous."
- "Did you hear the post office is thinking about charging seven cents just to mail a letter."
- "If they raise the minimum wage to $1.00, nobody will be able to hire outside help at the store."
- "When I first started driving, who would have thought gas would someday cost 25 cents a gallon. Guess we'd be better off leaving the car in the garage."

100 Vignettes

- "I'm afraid to send my kids to the movies any more. Ever since they let Clark Gable get by with saying "damn" in *Gone with the Wind*, it seems every new movie has either "hell" or "damn" in it."
- "I read the other day where some scientist thinks it's possible to put a man on the moon by the end of the century. They even have some fellows they call astronauts preparing for it down in Texas."
- "Did you see where some baseball player just signed a contract for $50,000 a year just to play ball? It wouldn't surprise me if someday they'll be making more than the President."
- "I never thought I'd see the day all our kitchen appliances would be electric. They are even making electric typewriters now."
- "It's too bad things are so tough nowadays. I see where a few married women have to work to make ends meet. It won't be long before young couples are going to have to hire someone to watch their kids so they both can work."
- "I'm afraid the Volkswagen car is going to open the door to a whole lot of foreign business."
- "Thank goodness I won't live to see the day when the Government takes half our income in taxes."
- "The drive-in restaurant is convenient in nice weather, but I seriously doubt they will ever catch on."
- "There is no sense going on short trips anymore for a weekend. It costs nearly $2.00 a night to stay in a hotel."
- "No one can afford to be sick anymore. At $15.00 a day in the hospital, it's too rich for my blood."
- "If they think I'll pay 30 cents for a haircut, forget it."

Those of you as old as me will remember polio in the 1940s and 1950s; 1955 was the year when Dr. Jonas Salk developed a vaccine for polio. Queen Elizabeth was the Monarch even way back then, while Louis St. Laurent was Prime Minister of Canada and W.A.C. Bennett was Premier of BC. We've gone through a few Prime Ministers and Premiers since then, but the Queen is still our official Head of State. January 7, 1955 was the date of the first TV broadcast of the opening of Parliament,

Only 56 Years Ago …

but this has never proved to be riveting TV. Dwight Eisenhower was the U.S. President; Winston Churchill was Prime Minister of Great Britain, while Nikita Khrushchev was the Russian leader.

Hockey fans may be interested in this: A riot erupted in Montreal on March 17, 1955 when Maurice Richard of the Montreal Canadiens was suspended, costing him the NHL scoring title and preventing him from participating in the playoffs. $50,000 was a lot of money in 1955, just to indicate how quickly things change; the first baseball player to earn over $1 million per year was Nolan Ryan in 1980.

The first McDonald's was erected in 1955 in the U.S. and fast foods and TV dinners appeared, including fish fingers. The first cans of Coca-Cola were sold—till then it had only been available in bottles. Rock-and-Roll music continued to grow in popularity with "idols" including Elvis Presley, Bill Haley and the Comets, Chuck Berry and The Platters, and many others.

VIGNETTE 73

Winter in the Lower Mainland vs. the Rest of Canada

I RECEIVED THE following note a few years ago from someone I had always thought of as a friend. "Chilled Vancouver commuters faced their second day of winter hell today as an additional ½ centimeter of peculiar white stuff fell, bringing BC's Lower Mainland to its knees and causing millions of dollars of damage. Local university scientists suspect that the substance is some form of water particles, and experts from Edmonton are being flown in for assistance. With the temperature dropping to almost zero, Vancouverites were warned to double insulate their lattes before venturing outside. Vancouver Police recommended that people stay inside except for emergencies such as running out of espresso or biscotti to see them through the most terrible storm to date. Canadian Tire reported that they had completely run out of fur-lined sandals. Drivers were cautioned to put their convertible tops up and some have been shocked to learn that their SUVs actually have four wheel drives, although most have no idea how to use it. Weary commuters faced soggy sushi and the threat of frozen breast implants even though the Coastal Health Authority assured everyone that they were safe, but down-filled bras were flying off the shelves at Mountain Equipment Co-op. One angry resident said that the government has to do something since they didn't pay $600,000 for a one bedroom condo to be forced into living like they were in Toronto."

Winter in the Lower Mainland vs. the Rest of Canada

Obviously the above was written by some jealous person living in a snowbound region of Canada! It is true that a couple of inches of snow can bring the Lower Mainland of BC to its knees, but the good news is that this snow usually disappears in a few days, and sometimes a few hours. It's also true that some of British Columbia's mountains get far more snow than anywhere else in Canada, but in the Lower Mainland we keep the snow mostly on the mountains. The highest average annual snowfall (1471 centimeters (48 feet) on Mount Fidelity) recorded was at a weather station near the Trans-Canada Highway on the west side of Glacier National Park. But let's look at the bigger picture. Parts of Newfoundland and Labrador, Quebec, and even Whistler, BC average more than 400 centimeters of snow each year. And even though many southern spots get more snow, it lingers longest in Canada's northern regions. The northernmost settlement, a military base at Alert, Nunavut on the northern tip of Ellesmere Island, is snow covered the longest, typically for 306 days a year.

Let's look at winter in Ontario where the Great Lakes have a huge impact on the amount of snow that falls in southern Ontario. Some snowbelt areas receive an annual average of well over 300 cm (120 inches) of snow annually. A fairly regular feature in southern Ontario, including Ottawa, is freezing rain that converts the roads into icy skating rinks. We lived in Ottawa years ago when the record for annual snowfall was set (which has since been broken). Just after you finish shoveling the driveway, the snowplow goes by and leaves a solid three foot bank that takes another hour to clear (only half an hour if you had a snow blower). The snow banks along our driveway were higher than the roof of our bungalow. Winter in Montreal is no picnic either, since it is exhibited by very cold, snowy, windy, and at times, icy weather, with an average high temperature of -5°C (23°F) and lows of -13°C (9°F) and an average annual snowfall of 218 cm (86 in). The lowest temperature ever recorded was −37.8 °C (−36 °F) on January 15, 1957.

Then there's Winnipeg, or Winter-peg as we love to call it. Due to its location in the Prairies, and its distance from both mountains and oceans, it has an extreme humid continental climate in that there are great differences between summer and winter temperatures. The openness of the prairies leaves Winnipeg exposed to numerous weather systems including cold Arctic high pressure systems. From December through February the maximum daily temperature exceeds 0 °C (32 °F), on average, for only 10 days and the minimum daily temperature falls below −20 °C (−4 °F) on 49 days. Cold weather and snow will occasionally extend into April. The coldest temperature ever recorded in Winnipeg was −47.8 °C (−54 °F), on December 24, 1879. (Summers are hot and humid, particularly in July, with frequent night time thunderstorms, with the maximum daily temperature exceeding 30 °C (86 °F) about 14 days a year). I was in Winnipeg some years ago in January when the temperature was −32 Celsius and a few minutes before we finished our meeting for the day my host used his remote start to start the car and get the heater going. A few months later (in July) he used the remote start again, only it was +33 so the air conditioning cooled the car down by the time we drove away. That kind of summarizes Winnipeg weather.

Calgary is also a city of extremes, and temperatures have ranged anywhere from a record low of −45 °C (−49 °F) in 1893 to a record high of 36 °C (97 °F) in 1919. Temperatures fall below −30 °C (−22 °F) on about five days per year. Calgary is said to be the only major city in the world where snowfall has occurred every month of the year, with the annual average being 126.7 cm (49.9 in). On the other hand, the city is among the sunniest in Canada, with 2,400 hours of annual sunshine, on average.

A discussion of Canadian winters would be incomplete without mentioning the Maritime Provinces. Who can forget the pictures on TV in 2011 of snowstorm after blizzard after snowstorm as people tried to dig themselves out? Snow blowers, and their operators, were gasping for breath. Snowstorms, which were usually accompanied by strong westerly winds with gusts up to 90 km/h, dumping more than a foot of snow each time, were a regular occurrence. Road conditions were regularly described as absolutely horrendous, and visibility was dramatically reduced on the Trans-Canada Highway near Fredericton, New Brunswick due to blowing and drifting snow. Motorists were regularly warned to 'proceed with caution.'

So let's get back to winter in Vancouver and the Lower Mainland. Give me my cold lattes any day! A friend from Ontario recently told me that he had actually worn out his snow shovel. Although popularly

Winter in the Lower Mainland vs. the Rest of Canada

thought of as being a rainy city, Vancouver has only 166 days per year with measurable precipitation, and 289 days with measurable sunshine. Nonetheless, from November until March, it is not uncommon for there to be 20 consecutive days with some amount of rain. The weather pattern known as the Pineapple Express often brings warm rainstorms in the winter. The annual average temperature in Vancouver is 10.1 °C (50.2 °F), with only Victoria being the major center that is warmer. The coldest month on record at Vancouver International Airport was January 1950, with an average high of –2.9 °C (27 °F).

While Victoria isn't in the Lower Mainland, we should give it honorable mention. Not many other Canadian cities (are there any?) can boast of doing a "Flower Count" in February. The 2011 count was 138,920,564, not including Butchart Gardens. While most of Canada outside of the Lower Mainland, and much of the U.S., is still in the cold clutches of winter weather, Victoria often enjoys spring (in February!) temperatures of 10-16° Celsius (up to 60° Fahrenheit). A few years ago, Victoria was said to have only one snow plow since their snow removal plan involved walking to the nearest coffee shop for a latte and waiting for the snow to melt in a day or two.

VIGNETTE 74

Everyone is Created Equal, But Some Are More Equal than Others

THE APRIL 2011 marriage of Prince William and Catherine Middleton sparked a lot of interest in the monarchy in Canada and elsewhere. Millions across the globe watched them tie the knot at Westminster Abbey, and many believe the couple brought renewed popularity to the Royal Family. The Royals had rather besmirched their reputation in recent years as playboys and playgirls and various ill-advised dalliances and divorces. Because of William and Kate's popularity, many people became interested in the issue of succession. Before changes were made in 2011, males came ahead of females in the succession list. That is, if William and Kate had produced a girl before 2011, and later a boy, then the son, even though younger, would have been heir to the throne when William was king. This, of course, was politically incorrect and various efforts, which included required approval by all Commonwealth countries, were made to change the law. If we accept the premise that males and females are equal, logic would suggest that all individuals are also equal, which begs the further question: why does anyone have an automatic inherent right to be Queen or King and be treated "royally" while others are treated as "commoners"? Just asking.

This leads to another question: is a constitutional monarchy a better system of government than a republic? A monarchy is a system of government under which a country is governed by one ruler, usually for his or her lifetime. Until the late 1700s, most countries were governed by monarchs who had titles such as emperor, king, prince, sultan or tsar. Today many countries of the world are republics, and a few are constitutional monarchies, where the real power is held by elected governments and where the monarch is head of state, but not head of the government. One example of a republic is the U.S., while Canada is a constitutional monarchy. In a republic a person is typically elected as Head of State with various titles such as president, and he or she has powers that are allowed to them by their constitution. In a monarchy a king/queen is a hereditary heir to the throne and has certain powers allowed by constitution. In a constitutional monarchy the monarch's powers are mostly cosmetic and their powers are carefully controlled by parliament. Although the U.K. has perhaps the best-known monarchy in the world, it not unique. Denmark, Sweden, Norway, the Netherlands, Belgium and Spain also function as constitutional monarchies, as do Japan and Thailand.

No British monarch has actually vetoed an Act of Parliament since 1720. But as unelected figures above the political conflicts of the day, they retain an important symbolic role as a focus for national unity (very important in Belgium, for example). In Britain their right "to advise, encourage and warn" the Prime Minister of the day has acted as a check against overly radical policies, and in Spain King Juan Carlos actually faced down a military coup in the 1980s. Queen Elizabeth II has reigned since 1952. She and her immediate family undertake various official, ceremonial and representational duties. As a constitutional monarch, the Queen is limited to non-partisan functions such as bestowing honors, dissolving Parliament and formally appointing the Prime Minister. Though the ultimate executive authority over the government of the United Kingdom is still by and through the monarch's royal prerogative, these powers may only be used according to laws enacted in Parliament, and, in practice, within the constraints of convention and precedent.

The British monarchy traces its origins from the Kings of the Angles and the early Scottish Kings. The last Anglo-Saxon monarch (Harold II) was defeated and killed in the Norman invasion of 1066, and the English monarchy passed to the Norman conquerors. In the thirteenth century, the principality of Wales was absorbed by England, and Magna Carta began the process of reducing the political powers of the monarch. In 1707, the kingdoms of England and Scotland were merged to create the

Kingdom of Great Britain and, in 1801, the Kingdom of Ireland joined to create the United Kingdom of Great Britain and Ireland.

The Queen's personal fortune has been the subject of speculation for many years. Sometimes estimated at U.S. $10–20 billion, recently Forbes magazine conservatively estimated her fortune at around U.S. $500 million (£280 million), not including some properties. This figure conflicts with a total addition of the Queen's personal holdings, for example, her personal art collection is worth at least £10 billion, but is held in trust for the nation, and cannot be sold. The Queen also owns large amounts of property privately that have never been valued, including Sandringham House and Balmoral Castle. The Queen also technically owns the Crown Estate with holdings of £6 billion; however, the income of this is transferred to the Treasury in return for the civil list payments. For me the question is, does the Queen and her family "own" these properties and art collections and bank accounts, or are they owned by the British people?

Does Canada pay to maintain the monarchy? Well, yes, and no. Canada is only responsible for the costs incurred by the royal clan while they are in Canada. Such a deal—I wish my vacations were paid for by the country that I'm visiting! But I guess that "state visits" aren't all vacation. Canada also pays for the cost of running the offices of the Governor General and our 10 provincial lieutenant-governors, which has more than doubled to $40–$50 million a year.

VIGNETTE 75

Advertising

WHY IS IT that grocery stores place headline grabbing magazines in a shopper's line of sight (and place piles of candy in front of their kids) at the checkout line? Answer: because this prompts people to buy stuff that they otherwise would not purchase. Advertising dates back to ancient times, with slogans and art painted on building walls in Roman times. However, advertising as we know it today was invented by Francis Ayer in 1869, whose agency was best known for creating iconic advertising slogans, such as "A Diamond Is Forever," "Be All You Can Be," "We Hear You," and many others. Advertising slogans are an effective way to promote a product or service and it is important to create effective and provocative customized slogans for a business. Typically slogans summarize claims about being the best quality, the tastiest, cheapest, most nutritious, providing an important benefit or solution, or being most suitable for the potential customer. "Let your fingers do the walking" (Yellow Pages), and "Leave the driving to us" (Greyhound) were other popular slogans a few years ago.

One of the earlier forms of advertising in the U.S. was on barn roofs with white paint. When the first cars rolled out of the Ford factory, America took to the road and farmers saw opportunity to sell fresh farm food to those who were passing by. Advertising created demand for new products, thereby changing the buying habits of Americans. Modern advertising developed with the rise of mass production in the late 19th and early 20th centuries. In 2010, spending on advertising was estimated at hundreds of billions of dollars worldwide.

Advertising is a form of communication used to persuade an audience (viewers, readers or listeners) to take some action with respect to products, ideas, or services. Most commonly, the desired result is to have the consumer buy something, but political and ideological advertising is now also common. Advertising messages are usually paid for by sponsors and viewed via various media, including traditional media—newspapers, magazines, television, radio, outdoor or direct mail—or new media such as websites and text messages. Non-commercial advertisers who spend money to advertise items other than a consumer product or service include political parties, interest groups, religious organizations and governmental agencies. Nonprofit organizations may rely on free modes of persuasion, such as public service announcements.

Commercial advertisers often seek to generate increased consumption of their products or services through branding, which involves the repetition of an image or product name in an effort to associate certain qualities with the brand in the minds of consumers. Branding is particularly evident in movies and in TV programs. It's not coincidental that the actors drive a particular model of car or have a can of Pepsi or a specific brand of coffee on the table. Companies have also used branding of sports stadiums providing rich contracts for cities or the owners, but this has periodically caused some embarrassment when the company suffers an economic reversal or changes owners. One stadium has had its name changed at least 3 times.

During the 1950s American businessmen began to suspect that consumers couldn't be "trusted" to know what products they wanted to buy, so of course they were willing to provide assistance. Since U.S. companies produced everything from cars to catsup in increasing amounts, it was in their interest to stimulate demand—that is, to convince consumers that they wanted or needed their products. A lot of advertising at this time was directed to "Mom the homemaker" in promoting laundry detergent or groceries or sewing machines, and many other products.

One example where advertising played a significant role is dieting, which became a major national issue in the 1950s, and the consumption of low-calorie soft drinks multiplied 300 times in a few years. The market was flooded with dietary aids, all of which were advertised with great gusto, even though a U.S. congressional committee found that nearly all of the dietary products were practically worthless, and that the American public had been paying over $100 million a year for phony latter-day patent medicines. Another example is trading stamps, which became a perfect advertising triumph even though the concept was clearly absurd and disadvantageous to the customer. While customers probably knew they were paying more in stores that gave stamps, the stamps nevertheless instilled a feeling of thriftiness. After all, the customers were getting something for "nothing" and, significantly, they were acquiring luxury items which they wouldn't ordinarily have bought. Housewives marched on U.S. State legislatures when their government was considering out-lawing trading stamps.

How is it that testimonials by local or national celebrities are so popular (and by extension so effective)? It seems apparent that the primary reason someone promotes a product is that they are being paid to do so, probably both in cash and free use of the product. In politics or business we would call that a conflict of interest, but in advertising it seems OK. Sure, some product hucksters believe and actually use the product they are promoting, but the fact that they gain from promoting the product detracts from the meaningfulness of their message.

Web advertising has grown and changed considerably since its beginnings in the early 1990s. Online ads offer customers a more interactive way to shop and buy at the same time they give sellers additional ways of reaching the public. Buying things online and providing personal information including credit card numbers also increases the possibility identity theft and of buyers being scammed. In 2001, online ad revenue exceeded print ad revenue for the first time. This took a while, with online ad revenue totaling $86 million in 2001, but surpassing $2.2 billion in 2011. TV ads still garner the most revenue, and 24 percent of the total at $3.4 billion, but the rate is slower than for online ad revenue, which in 2010 had 16 percent of the total compared to 15 percent for daily newspapers. Advertising must work if companies are prepared to spend so much money on it!

VIGNETTE 76

TOMS Shoes, One for One, and Mossy Foot

I FIRST HEARD of TOMS Shoes a few years ago. The TOMS Shoes company was started by Blake Mycoskie in 2006 after he visited Argentina and noticed how many children did not have any shoes. His dream for this company was to create a shoe that he could both sell *and* give away. He wanted his company to make it possible to donate one pair of shoes to a child in need with every pair purchased. To realize this mission, Mycoskie made a commitment to match every pair of TOMS purchased with a pair of new shoes to send to a child in need, called One for One. "I was so overwhelmed by the spirit of the South American people, especially those who had so little," Mycoskie said. "And I was instantly struck with the desire—the responsibility—to do more."

He recognized the traditional Argentine alpargata shoe as a simple, yet revolutionary solution. He quickly set out to reinvent the alpargata for the U.S. market with a simple goal: to show how together people can create a better tomorrow by taking compassionate action today. During its first year in business, TOMS sold 10,000 pairs of shoes. Mycoskie returned to Argentina the following year with family and friends and gave back to the children who had first inspired him. Thanks to supporters, TOMS had given over one million pair of new shoes to a child in need as of September 2010. TOMS now gives away shoes in over 20 countries and works with charitable partners in the field who incorporate shoes into their health, education, hygiene, and community development programs.

TOMS Shoes, One for One, and Mossy Foot

TOMS' giving partners are made up of NGOs (Non-Governmental Organizations), charities, and non-profits already established and working in the countries in which TOMS gives shoes. Their expertise guides TOMS to give new shoes responsibly, making sure there aren't adverse socioeconomic effects, and to ensure that sustainable giving is possible. Giving shoes to the same children on a regular basis is the idea upon which TOMS was started, and is what truly improves the lives of children and their communities.

It didn't take long for the world to notice this new approach to business—in 2007 TOMS Shoes was honored with the prestigious People's Design Award from the Cooper-Hewitt National Design Museum, Smithsonian Institution. Two years later, Mycoskie and TOMS received the 2009 ACE award from Secretary of State Hillary Clinton, which recognizes the company's commitments to corporate social responsibility, innovation, exemplary practices, and democratic values worldwide. The TOMS mission of giving shoes away has attracted other brands, resulting in unique and successful collaborations. Ralph Lauren sold co-branded Polo Rugby TOMS, giving away a matched pair with every pair sold. Element Skateboards has issued limited edition TOMS + Element shoes as well as a One for One skateboard. With every skateboard purchased, one will be given to a child at the Indigo Skate Camp in Durban, South Africa. It is TOMS' hope that as the One for One movement continues to grow, more and more companies will look to incorporate giving into what they do.

Apparently huge numbers of people are wearing TOMS, which are said to be almost as popular as flip-flops. And they are fashionable. I didn't find any retail outlets in BC that sell TOMS, but check out the TOMS web site and help the cause!

Another innovative TOMS idea was to organize groups of supportive Americans to spend a day walking without shoes to raise awareness for those who don't have any shoes. Wearing shoes also prevents feet from getting cuts and sores. Not only are these injuries painful, they also are dangerous when wounds become infected. Mycoskie also has a goal of stamping out foot diseases that can spread from the ground to bare feet and released a product in 2011 that will help do this. At one point TOMS was sending 10,000 pairs a month of specially developed shoes to Ethiopia, particularly for the estimated 1 million impoverished people suffering from a terrible foot disease called podoconiosis, which is on the WHO list of neglected tropical diseases.

Podoconiosis is also called "podo" or "mossy foot", and the technical description is a "profuse velvety papillomatous growth that develops large warty projections caused by chronic lymphedema and stasis with maceration and associated infection of the feet." Another description is a non-infectious disease causing massive swelling of the legs leading to pain, disability and social exclusion. Podo is grotesque. In severe cases, the victim's feet appear to be turning into a cauliflower—horrible, rotting cauliflower—like something that grows under a rock in 20 feet of water. These are nightmare feet, seeming to bubble and melt, producing unbearable odors. An estimated one million Ethiopians suffer from podo, as do perhaps three million other Africans. In affected areas—typically mountains with red volcanic soil—1 out of every 20 people may have it. The cause of podo is related to Ethiopia's volcanic soil, since tiny silica crystals become embedded in victims' lymphatic tissue. He found that the silica made its way through the skin of feet and ended up scarring or blocking lymphatic channels, causing swelling and deformity. Dr. Nathan Barlow later founded the Mossy Foot Project, which is headquartered in southern Ethiopia. Besides its focus on care and prevention, Mossy Foot gives away shoes specially made in its workshop, since many people with podo cannot wear standard shoes. Mossy Foot distributes supplies like soap and bleach, shoes for children of podo victims, and also offers job training and microenterprise loans for patients to support themselves. TOMS Shoes also plays a key role since a pair of shoes can change a person's life. TOMS targets its shoe distribution to the children of podo patients, who are five times more likely than the general population to get podo.

VIGNETTE 77

Why Can't Europeans Be More Like Us?

IF YOU ARE of a certain advanced vintage, or if you like old fashioned musicals, you probably remember Rex Harrison's semi-rant semi-rap in "My Fair Lady" plaintively asking the question, "Why can't a woman be more like a man?" We recently toured Europe by bus, staying in a variety of three star hotels, and I observed a great many differences between Europe and North America. My question is, "Why can't Europeans be more like us?!"

Let's start with the hotels; why can't they be more like the ones in Canada? Why do they number the floors differently? You don't walk in on the first floor; you walk in on floor "0" and have to walk one floor up to "floor #1". This causes no end of confusion when ordinary tourists come down on the elevator and get off on the "first" floor and then can't find the lobby. And they call them "lifts," not elevators. When you walk into your hotel room, the light switches are backwards—up and down are opposite to what we are used to. Next, none of the lights work. Well they don't work until you place your electronic key card into a little holder by the door. Actually this is a pretty good idea—when you take your card with you as you leave the room, everything is shut off a few minutes later, which saves on electricity. (A few hotels still had the old fashioned giant sized keys like in a Humphrey Bogart movie that you had to turn in at the reception desk whenever you left the hotel. And in some cases you deposited your room key into its numbered slot beside the front desk which then gave you a key to the front door; then you reversed the process when you came back to the hotel). And why is it they use 220 volts rather than 110? Perhaps it is to sell adaptors and replace blown up hair dryers.

The most obvious difference in the hotel rooms was that most of them were equipped with two single beds, side by each. Sometimes the beds were separated by a night table, but in every case each bed of course had its own bedding. And this bedding didn't include a sheet with which to cover yourself in addition to a blanket. Instead, they have a sheet that buttons around a heavy quilt so one typically either roasts or freezes when you finally throw off the heavy cover in the middle of the night. This leads us to the air conditioning, which was great—you just open the window, which was fine for us since it was May and fairly cool at night. It seemed interesting that some of the hotel rooms had hardwood floors; the buildings must be well constructed, since we didn't hear people clomping around in the rooms above us.

The bathrooms are a whole separate issue. Some of them were designed by midgets or contortionists. The showers were fascinating; some were just walk-in, with a side glass panel but without a door—which actually worked OK. Others had the shower in the deep, deep tub that sometimes seemed to have a specially designed slippery bottom. Usually the removable shower head on the end of a flexible hose had a holder on the wall so you could have both hands free, but in one hotel there was no wall holder. This meant that you either needed a personal assistant to help you take your shower or you used a more extended process in getting wet, then using shampoo and soap and then picking up the shower hose and head, hoping that it didn't wildly flop around and get the whole bathroom showered when you turned on the water. They don't seem to believe in wash cloths but they do have towels. Public washrooms are different. Almost everywhere you need to pay to use the washrooms; sometimes there is an attendant, and in other cases there is a turn-style. But the washrooms are always clean.

Of course, we could learn some things from the Europeans. Many buildings in Europe are very old but still in excellent shape. We seem to tear down buildings after they reach an advanced age like 50, but we saw many buildings that were more than 300 years old and well maintained and functional. Many buildings have real shutters for safety or to be used as blinds, not the fake decorative ones we often have here.

Why Can't Europeans Be More Like Us?

The shutters can be opened or closed manually, or some are operated electronically. Most medium and large cities have street cars, which I always found fascinating, and of course the large cities have extensive subway systems, called the tube or the underground. The cathedrals were inspiring and depressing—inspiring because of their massiveness and beautiful architecture and their ancientness, but discouraging because tourism seemed to be more important than worship.

Perhaps Europeans enjoy life more than we do. They stay up late (in some countries restaurants don't open until 7 or 8 pm) and get up late, and in some cities they have a two hour siesta in the afternoon when most of the shops close. Two great features are the sidewalk cafes and the city squares where people can watch the world go by. There are millions of small shops and not so many shopping malls. I'm not sure that they enjoy driving—except for red lights, there don't seem to be any rules other than making sure their horns work as they aggressively careen around the streets that often don't have any lanes, or at least lanes that mean anything. The cars are smaller but they drive faster. Except for a few main streets, the streets are very narrow. The apartments or condos, called "flats," come without parking so people may need to parks blocks away from their place. Gas stations are generally unobtrusive; they just look like bus stops where you pull up for a fill.

But visiting Europe is a fabulous experience, and a person needs to respect their practices and traditions.

VIGNETTE 78

Here's a Hot Topic

FIRE HAS BEEN known to man since the very earliest times. In certain caves in Europe in which men lived thousands of years ago, charcoal and charred bits of bone have been found among stones that were evidently used as fireplaces. We can only guess how men learned the trick of making a fire; in fact they probably knew how to use fire before they knew how to start it. For example, lightning might strike a rotten tree and the trunk would smolder and they would light a fire and keep it going.

The ancient Greeks and Romans used a kind of lens, called "a burning glass," to focus the rays of the sun to produce enough heat to ignite combustible material. In China and India, a piece of broken pottery was struck against a bamboo stick. The outer coating of the bamboo is very hard and has similar qualities as flint. Indian tribes used to rub sulphur over two stones and strike them together, and when the sulphur ignited they would drop the burning stone on dried grass or other material. Eskimos used pieces of quartz against a piece of iron pyrite. Among the North American Indians, rubbing two sticks together to produce fire was a common method.

Three things are needed to start a fire, namely fuel, heat, and oxygen. Heat breaks down long cellulose molecules into shorter molecules that can react with oxygen in a process called 'pyrolysis.' Smaller pieces of tinder or kindling will heat up more easily than larger pieces. Heat can be suppressed with dirt or water, by spreading the fuel apart, or by adding green or wet wood, which requires more heat to dry and ignite it.

Arson is the crime of wilfully or maliciously setting a fire for unlawful or improper purposes. Arsonists set fires for a multitude of reasons, including vandalism, revenge, monetary gain, and mental illness. Whether used to cover up a crime, or as a violent act against another person's property, arson is a destructive method of achieving unethical goals. Beyond that, it carries the risk of severe injury, if not loss of life, to others. Fire investigation has become a forensic science, seeking to determine a fire's origin and cause. During a fire investigation, the investigator tries to uncover the source and path of the fire, using clues such as burn patterns. Arson investigators rely on knowledge of the "behavior of fire," the basic "fire triangle" of heat, oxygen, and fuel, the way the surrounding environment affects fire, and the different modes of fire to determine whether arson is implicated. Unfortunately, arson is one of the most difficult crimes to solve, and arrests are only made in less than 25 percent of suspected arson cases.

If an accelerant is used to start a fire, a small amount will likely still be present in the charred debris. The presence of an accelerant (a chemical "fuel" that causes a fire to burn hotter, spread more quickly than usual, or be unusually difficult to extinguish) or ignitable liquid such as gasoline or another petroleum distillate can indicate an incendiary fire or arson. Areas suspected to contain ignitable liquids are collected and sent to forensic laboratories to be examined, using techniques such as gas-liquid chromatography and mass spectrometry. Various techniques exist to detect accelerants at a fire scene. These range from experienced fire investigators or specially trained "sniffer" dogs using their sense of smell to detect the characteristic odor of various accelerants in the surrounding air, to more complex laboratory methods.

The most commonly used accelerants are gasoline, kerosene, turpentine and diesel fuel. These accelerants are generally complex mixtures of hydrocarbons which have similar chemical properties, but their boiling points vary and this variation causes the accelerants to alter their composition during the evaporation or boiling process. Various analytical techniques take advantage of these features to detect the presence of an accelerant amongst the debris after an arson fire. As accelerants evaporate, the hydrocarbons move into the air above the fire debris, which is called the "headspace." During the evaporation process, the headspace above the accelerant becomes concentrated with the more volatile hydrocarbons and so has a different composition than the accelerant left in the debris. It is this headspace which is tested by the various techniques to detect the presence of an accelerant in the debris remaining after a fire.

An important part of the investigator's job is to correctly collect evidence from the scene to further the investigation. The samples collected by a fire investigator will be analyzed in a laboratory for the presence of any chemicals which could have been used as accelerants. Gas chromatography involves separating mixtures of chemicals or gases into their individual components based on the different boiling points of their hydrocarbons. Each gas or component in the mixture can then be identified, because each produces a distinct chemical fingerprint called a chromatogram. In headspace gas chromatography, solid debris taken from the suspected point of origin of the fire is placed in an airtight vial to prevent any accelerants from evaporating. The vial is then heated in the laboratory, releasing the accelerant's hydrocarbons into the trapped headspace above the debris. A needle is inserted through the cap of the vial to remove a sample of the hydrocarbons and inject them into a gas chromatograph instrument for separation and analysis.

VIGNETTE 79

The "Machinery of Government" and Other Things You May Not Have Heard About

WHEN I WORKED for the federal government I once attended a training course where one of the attendees was from a unit called "The Machinery of Government" (MOG). He was called away during the course, since there had been a cabinet shuffle and his office had some procedures that needed to be implemented. I found out that the MOG deals with the interconnected structures and processes of government, such as the functions and accountability of departments in the executive branch of government. The term MOG is used particularly in the context of addressing changes in administration where different elements of government "machinery" are created.

The MOG Secretariat in Ottawa provides advice and support to the Prime Minister and the Clerk of the Privy Council on the structure and functioning of the government as a whole. It's easy for wiseacres and those blessed with a spirit of criticism to say that the government "machinery" is broken, but we should look more closely at how this machinery is supposed to work. Much of their work involves supplying advice on the responsibilities of the Prime Minister, including ministerial mandates; the structure of Cabinet, its committees and their decision-making processes; the structure and organization of government portfolios; the roles and accountability of senior public office holders; and helping manage transitions from government to government. We can see that this unit is particularly busy when the government changes after an election or even after a government is re-elected, but there are also on-going responsibilities.

The primary responsibility of the Privy Council Office (PCO) is to provide non-partisan, public service support to the Prime Minister and Cabinet and its decision-making structures to facilitate the smooth and effective operation of the Government on policy and operational issues. Led by the Clerk of the Privy Council, the PCO helps the Government implement its vision and respond effectively and quickly to issues facing the government and the country. The PCO is the Prime Minister's "Department" as well as the Cabinet Secretariat and is staffed by career public servants. In today's world of complex economic and social issues, much of this work involves consultation, coordination and collaboration both inside and beyond the federal government. The PCO also works to achieve high professional and ethical standards in the federal Public Service.

To provide more detail, the PCO has three main roles:

(1) Advisor to the Prime Minister. PCO brings together quality, objective policy advice and information to support the Prime Minister and Cabinet. This includes providing advice and information from across the Public Service; consultation and collaboration with international and domestic sources inside and outside government (including the provinces and territories); and information on the priorities of Canadians.

(2) Secretary to the Cabinet. PCO facilitates the smooth, efficient and effective functioning of the federal government Cabinet and the Government of Canada on a day-to-day basis. This includes management of the Cabinet's decision-making system; coordination of departmental policy proposals to Cabinet (with supporting policy analysis); scheduling and support services for meetings of Cabinet and Cabinet committees; advancing the Government's agenda across federal departments and agencies and with external stakeholders; advice on government structure and organization; preparation of Orders-in-Council and other tools to give implement Government decisions; and administrative services to the Prime Minister's Office.

(3) Public Service Leadership. The PCO ensures that Canadians are served by a quality public service that delivers advice and services in a professional manner, and strives to meet the highest standards of

The "Machinery of Government" and Other Things You May Not Have Heard About

accountability, transparency and efficiency. This includes management of the appointments process for senior positions in federal departments, Crown Corporations and agencies; setting policy on human resources issues and public service renewal; ensuring the Public Service has the capacity to meet emerging challenges and the changing responsibilities of government; and submitting an annual report to the Prime Minister on the state of the Public Service.

The Treasury Board is a Cabinet committee established by law and composed of several ministers responsible for the management of government expenditures and human resources in the public service. The Treasury Board Secretariat supports the Treasury Board in these responsibilities. While the Department of Finance is responsible for establishing general policy on government revenues and expenditures, the Treasury Board oversees the management of the budget and plays a co-ordinating role in the preparation of the budget. According to the *Financial Administration Act*, the Treasury Board can deal with any questions concerning financial management, giving it authority over departmental budgets, expenditures, financial commitments, revenue, accounts, personnel management and all the principles governing the administration of the public service. Basically, the Treasury Board is the employer, expenditure authority and general manager of the public service.

The late Senator Eugene A. Forsey originally wrote *How Canadians Govern Themselves* to answer questions that people may have about Canada's system of government. We should all check it out so that we can more effectively support and criticize our government.

VIGNETTE 80

Is the United Nations Effective?

I RECALL LEARNING about the United Nations (UN) in Social Studies in high school and the UN then seemed to me to be a mysterious organization. The objective of the UN is to facilitate cooperation in international law, international security, economic development, social progress, human rights, and achievement of world peace. The UN was founded in 1945 after World War II to replace the League of Nations to stop wars between countries, and to provide a platform for dialogue. Some critics say that the United Nations is a congress of quarrelers and self-seekers. For various reasons the League of Nations did not prove to be very effective in the earlier part of the 20th century, and after World War II the UN was formed to promote international peace. Eventually it established its headquarters in New York City. Is the UN any more effective than the League of Nations? Many say no, and there is a lot of evidence to support this contention even though it also does some good work.

There are currently 192 member states, including every internationally recognized country in the world except Vatican City. The UN and its specialized agencies decide on substantive and administrative issues in regular meetings held throughout the year. The organization has five principal operating sections: the General Assembly (the main deliberating assembly); the Security Council (for deciding certain resolutions for peace and security); the Economic and Social Council (for

Is the United Nations Effective?

assisting in promoting international economic and social cooperation and development); the Secretariat (for providing studies, information, and facilities needed by the UN); and the International Court of Justice (the primary judicial body).

Other prominent UN agencies include the World Health Organization (WHO), the World Food Program (WFP) and United Nations Children's Fund (UNICEF). The UN's most visible public figure is the Secretary-General. As of spring 2011 this was Ban Ki-Moon of South Korea. The organization is financed from assessed and voluntary contributions from its member states, and has six official languages: Arabic, Chinese, English, French, Russian, and Spanish.

There has consistently been controversy and criticism of the UN. The cynics say that the UN can't even get rid of pirates in little dingys off the coast of East Africa, so the likelihood of solving serious world problems isn't high. In most cases the UN can't effectively threaten or influence specific nations because they can't agree on how to carry out any threat, nor agree on economic sanctions, since some countries often work around the sanctions and still buy or sell from the problem nations. Each of the "permanent" members of the Security Council has veto power, so that many potentially effective actions are thwarted.

One of the bigger UN scandals was the oil-for-food program which began as a humanitarian plan in 1995 to soften the sanctions against Saddam Hussein's Iraq. The plan was for the UN to allow Iraq to sell a certain amount of oil each year, provided that the profits were used to buy food, medicine and other necessities for Iraq's people. Instead, the oil was allotted to politically connected insiders, allegedly including the head of the oil program himself. The insiders got rich while Iraq's people continued to suffer. One beneficiary of the program was the UN Secretary-General's own son, who was employed by a firm that played a key role in the scandal-plagued program. Another major UN embarrassment was the "Durban Declaration"—a blatantly anti-Semitic diatribe against Israel at the "World Conference Against Racism, Racial Discrimination, Xenophobia and Related Intolerance" in 2001. Canada, followed by the U.S., Israel and a few European countries, walked out of the conference since the conference report was the exact opposite of the conference objectives.

The promotion and protection of human rights has been a major preoccupation for the United Nations. The General Assembly declared that respect for human rights and human dignity "is the foundation of freedom, justice and peace in the world" in the Universal Declaration of Human Rights. Over the years, a whole network of human rights instruments and mechanisms has been developed to ensure the primacy of human rights and to confront human rights violations. One criticism was that the Commission did not engage in constructive discussion of human rights issues, but instead was a forum for politically selective finger-pointing and criticism. The Commission was repeatedly criticized for the composition of its membership, since several of its member countries themselves had dubious human rights records. Such states saw membership in the Commission as a way to defend their poor human rights records—countries such as China, Zimbabwe, Russia, Saudi Arabia, Pakistan, Algeria, Syria, Libya, Uganda and Vietnam all had extensive records of human rights violations. One concern was that by working against resolutions on the commission condemning human rights violations, these countries indirectly promoted despotism and domestic repression. At one point Sudan became a member of the commission, which many countries saw as absurd since Sudan was at that time conducting ethnic cleansing in its Darfur region and as a result some countries were unwilling to work through the commission. The UN's bureaucratic operating rules were built on the idea, shall we say "ideal" that every nation would be honorable and as such the UN was not really able to deal with issues when this inevitably proved not to be true.

Perhaps the best that can be said of the UN is that it serves as a forum for open dialogue between nations, and this alone may be a powerful enough argument for nations to continue supporting the organization.

VIGNETTE 81

Country Music

GROWING UP IN Alberta meant that we were bombarded daily with country and western music, which was OK with me. I still listen to this type of music on occasion, much to the consternation of my family. In any case, I wanted to check out the history of this music genre. Though musicians have been recording their fiddle tunes for quite some time, true country music as a blend of traditional and popular musical forms didn't get its start until the 1920's. Around this time, folk music became slightly more sophisticated and the Grand Ole Opry radio station opened in Nashville, Tennessee. The term *country music* gained popularity in the 1940s when the earlier term *hillbilly music* came to be seen as denigrating.

Immigrants to the Maritime Provinces of Canada and the Southern Appalachian Mountains of the U.S. brought the music and instruments of the Old World along with them for nearly 300 years. They brought some of their most important valuables with them, and to most of them this was a music instrument. Early Scottish settlers enjoyed the fiddle because it could be played to sound sad and mournful or bright and bouncy. Outside of the U.S., Canada has the largest country music fan and artist base. Mainstream country music is culturally ingrained in the Maritimes and the Prairie Provinces, areas with large numbers of rural residents. Country music is popular in the United States, Australia, Ireland, New Zealand, Canada, and the United Kingdom. The sounds and style of country music have changed drastically over the years and today there are many different variations of country music including blue grass, honky tonk, western-country, rockabilly, and many more.

Country music is a mystery to everyone who's not a fan. "Un-fans" and critics say that the themes and issues range from drinking, cheating, nostalgia, class consciousness, truck driving, love, unrequited love, death, loneliness, Christianity, patriotism, rural life, family, prison, outlaws, cowboys, murder, loss, courting, misery, morality, trains, wanderlust, poverty, to self-destruction and redemption. Truck driving country music is a special genre of country music, and has the tempo of country-rock and the emotion of honky-tonk, and its lyrics focus on a truck driver's lifestyle. One consistent theme in modern country music is that of proud, stubborn individualism.

One effect of the Great Depression in the 1930s was to reduce the number of records (for you young folks, these were the round disc things that were used before CDs were invented) that could be sold. Radio became a popular source of entertainment, and "barn dance" shows featuring country music were started all over the South, as far north as Chicago, and as far west as California. The most important show was the *Grand Ole Opry*, aired starting in 1925 by WSM-AM in Nashville, and it continues to the present day.

Many musicians performed and recorded songs in any number of styles. Bill Haley sang cowboy songs, and was at one time a cowboy yodeler, but became most famous as an early player of rock n' roll, adding Jimmie Rodgers-stylings to his environment, thus creating a sound that was very much his own. In the late 1940s, country crooner Eddy Arnold placed eight songs in the top 10 hit parade. By the end of World War II, "mountaineer" string band music known as bluegrass had emerged when Bill Monroe joined with Lester Flatt and Earl Scruggs, introduced by Roy Acuff at the Grand Ole Opry. Gospel music, too, remained a popular component of country music. Red Foley, the biggest country star following World War II, had one of the first million-selling gospel hits ("Peace in the Valley") and also sang boogie, blues and rockabilly.

Another type of stripped down and raw music with a variety of moods and a basic ensemble of guitar, bass, steel guitar, and later drums became popular, especially among poor white southerners. It became known as honky tonk and had its roots in Texas. Rockabilly was most popular with country fans in the 1950s; this was a mixture of rock-and-roll and hillbilly music. Elvis Presley partially converted over to country music, and he played a huge role in the music industry during this time. The number two, three and four songs on *Billboard's* charts for that year were Elvis Presley's "Heartbreak Hotel"; Johnny Cash's "I Walk the Line"; and Carl Perkins' "Blue Suede Shoes." Beginning in the mid-1950s, and reaching its peak during the early 1960s, the Nashville sound turned country music into a multimillion-dollar industry.

In 2005, country singer Carrie Underwood rose to fame as the winner of the fourth season of *American Idol.* Underwood also made history by becoming the seventh woman to win Entertainer of the Year for the Academy of Country Music Awards,

From Johnny Cash's "Every Time I Itch I Wind Up Scratching You" to Kenny Chesney's "She Thinks My Tractor's Sexy," country music song titles have never ceased to surprise people with their quirkiness. Those making fun of country music remind us of song titles like: "I'm So Miserable Without You; It's Almost Like Having You Here," "If the Phone Doesn't Ring, You'll Know It's Me;" "If You Won't Leave Me Alone I'll Find Someone Who Will;" "When You Leave, Walk Out Backwards so I'll Think You're Coming In;" and "I've Got Tears in My Ears Lying on My Back Crying over You."

VIGNETTE 82

Sweet Hummingbirds

MY SWEETIE ENJOYS watching hummingbirds as they flit and fly to her brightly colored sugar water feeder, even through the winter. Hummingbirds are among the smallest of birds, most species measuring about 10 cm. The Bee Hummingbird at 5 cm is the smallest bird known. They can hover in mid-air by rapidly flapping their wings 12–90 times per second and can also fly backwards—the only group of birds able to do so. Their name derives from the characteristic hum made by their rapid wing beats, and they can fly at speeds exceeding 50 kph or 30 mph. Hummingbird flight has been studied intensively from an aerodynamic perspective using wind tunnels and high-speed video cameras. The average hummingbird lifespan is 3 to 5 years.

Hummingbirds thrive on nectar, a sweet liquid inside certain flowers; in fact they consume more than their body weight in nectar each day and to do so they must visit hundreds of flowers daily. Like bees, they are able to assess the amount of sugar in the nectar they eat; they reject flower types that produce nectar that has less than 10 percent sugar content. Nectar can be a relatively poor source of nutrients, so hummingbirds also meet their needs for protein and other nutrients by also preying on insects. Hummingbirds have a special bill that opens only slightly and allows their tongue to dart out and into the interior of flowers. Because of artificial feeders and winter-blooming gardens, hummingbirds sometimes stay in northern climates all year. Bird feeders, using sugar water, allow people to observe and enjoy hummingbirds up close while providing the birds with a reliable source of energy, especially when flower blossoms are less abundant. Boiling and then cooling this mixture before use deters the growth of bacteria and yeasts. Hummingbirds may return to the same gardens each year.

To conserve energy, hummingbirds spend an average of 10–15 percent of their time feeding and 75–80 percent sitting and digesting. Many plants pollinated by hummingbirds produce flowers in shades of red, orange, and bright pink, though the birds will take nectar from flowers of many colors. With the exception of insects, hummingbirds while in flight have the highest metabolism of all animals, a necessity in order to support the rapid beating of their wings. Their heart rate can reach over 1,200 beats per minute. Hummingbirds are continuously hours away from starving to death, and are able to store just enough energy to survive overnight. Hummingbirds can dramatically decrease their metabolism rate and heart rate at night when food is not readily available. In contrast, somehow hummingbirds can travel 800 km (such as across the Gulf of Mexico) on a nonstop flight. They prepare for such a flight by storing up fat to serve as fuel, thereby augmenting their weight by as much as 100 percent to increase their flying time. Hummingbirds are found in the Americas from southern Alaska to the tip of South America and in the Caribbean.

Hummingbirds usually build a cup-shaped walnut-sized nest on a tree branch. The nest varies in size depending on the species. In many hummingbird species, spider silk is used to bind the nest material together and secure the structure to its support. The unique properties of silk allow the nest to expand as the young grow. Two white eggs are laid, which, despite being the smallest of all bird eggs (about 'Tic-Tac' size), are in fact large relative to the hummingbird's small adult size. The mother feeds the nestlings on small insects and nectar by inserting her bill into the open mouth of the nestling.

Keeping bees, yellow jackets and ants away from the hummingbird feeders is a challenge. Bees seem to be attracted to the yellow part of a flower so red bird feeders work better and reduce or eliminate the bee and ant problem. Mint extract painted near the feeding ports may also work. Flowers are pollinated in a number of ways, including the primary assistance of bees, but also by hummingbirds, wasps and butterflies that all feed on the flower pollen and facilitate its spread. Without this help, flowers would have a hard time spreading their pollen and reproducing.

VIGNETTE 83

Conspiracy Theories and Theorists

A CONSPIRACY THEORY is a theory that explains a historical or current event as being caused by a plot by powerful and/or cunning conspirators to achieve their own purposes. Conspiracy theories should be viewed with skepticism because they are rarely supported by any conclusive evidence. Conspiracy theorists speculate on the motives and actions of "secretive" individuals. Rational explanations or criticisms are not accepted, and the conspiracy theory is not rational in that it may be impossible to prove it as wrong. A conspiracy theory generally is a closed system of ideas which explains away any contradictory evidence by claiming that the conspirators themselves planted it.

The following is a list of "popular" conspiracy theories as seen in Google and other sources:

1. 9–11 was an inside job. This one is so fantastic that it's hard to believe anyone could actually suggest this, but some folks believe that President Bush knew about it and that 9-11 was orchestrated by the CIA, or the Mafia, or Fidel Castro.
2. Lee Harvey Oswald did not act alone in the assignation of John F Kennedy, or he didn't even do it.
3. The death of Diana Princess of Wales was masterminded by the British Royal Family.
4. Elvis is working at a Wal-Mart in Alabama or…. There's a lot of emotion by Elvis fanatics to this theory since it's so hard to

believe that such a young man could die, but the passage of time means that the theory is gradually losing steam.
5. Osama bin Laden died several years ago, but the U.S. kept up the charade that he was alive for their own political purposes.
6. The CIA determines U.S. foreign policy, or the Mafia controls U.S. foreign policy, or big business corporations control U.S. policy or … How else does one explain all the bad things that happen in the government?
7. The Falun Gong is a secret society aiming to bring down the Chinese government. I guess this shows that even governments can believe in ill-founded conspiracy theories. The Chinese government is very suspicious of any group or individual that doesn't automatically obey every policy and pronouncement by the government. The opposite conspiracy theory is held by some wealthy Chinese, who believe that their government could at any time confiscate their assets, particularly if these citizens voice any criticism of the government. This is said to be one reason why many wealthy Chinese are buying property in Vancouver and other offshore locations just in case they need to flee China. In this case there may be some validity to the idea of this conspiracy!
8. Prime Minister Harper has a hidden agenda. Opposition parties and the media in Canada have been trumpeting this theory for several years, but we saw this dramatically demonstrated in the spring of 2011, when a lowly House of Commons Page risked her job to unfurl a banner in the House and loudly proclaim that Steven Harper's agenda needed to be stopped. Mr. Harper's "hidden agenda" hadn't been demonstrated in the previous years as Prime Minister, but I suppose that the 2011 Conservative majority re-ignited the conspiracy theory.
9. Global Warming is a product of the imagination of influential environmentalists. This may or may not be a conspiracy theory, but when a leading environmental agency was shown to have cooked their data the conspiracy theorists leapt into action.
10. Car manufacturers could build electric cars, or at least more efficient internal combustion engines, but the oil companies or the government or the automobile industry killed or delayed

their development. There may be some validity to this particular theory; you may have seen the documentary entitled "Who Killed the Electric Car," where three usual suspects were suggested—the American government since they needed to protect the revenue from oil and gas sales; the car manufacturers who destroyed hundreds of tested and effective electric cars to prevent their introduction into the market; and the oil companies who bought out (and then closed) a small company that had developed a lighter weight but effective battery.
11. The cancellation of the Canadian Avro Aero was dictated by the U.S. This plane was ahead of anything that had been designed to date and it baffled many Canadians when Prime Minister John Diefenbaker cancelled the program and let the U.S. industry take the lead.
12. Ontario and Toronto are the center of the Universe.
13. The NHL and Gary Bettman discriminate against Canadian hockey teams.
14. Gasoline prices are fixed.

All right, I admit it; I mention the last three issues to show that I'm not immune from the lack of objective thinking, but that doesn't mean they're not true. You have to admit that there is pretty strong evidence in each case. I once heard an interview with a petroleum industry association executive, and this fellow was so persuasive that he could have made Joseph Stalin sound like a choir boy.

I suppose one reason for the prevalence of conspiracy theories is the lack of trust that people have in authority. It's evident of course that this lack of trust is frequently highly justified. Officials often feel compelled to tell only part of the story in a given situation, or distort what happened, or in fact lie to protect themselves, or for some other reason. But it's still baffling why people have such difficulty in accepting objective evidence. Of course, conspiracy theorists will not accept any evidence, regardless of how objective or how well documented the evidence, because if the evidence is overwhelming, this just proves that this evidence has been "cooked" to fit the situation and the official explanation. But I think that there are other reasons as well, one being that our human nature doesn't want to face the truth, or we realize that we ourselves often act dishonourably. Another reason may be that we just love to be critical. Some actually make good money spreading these thoughts through society. Others simply use conspiracy theories to solidify their hatred towards an individual or group.

VIGNETTE 84

BC—Why Would You Live or Vacation Anywhere Else?

"BEAUTIFUL BC" IS more than just a slogan; there are many superlatives that we can use to describe British Columbia; one is "Super, Natural". There are many dramatically beautiful areas. Picture in your mind one or two of your favorite places in BC—a place where you would like to visit for an hour, or a whole day, or two weeks, or in fact where you would actually like to reside. BC in general can be said to be one of the best places in the world in which to live. Let's take a whirlwind tour of BC and see if I mention your favorite spot.

Grouse Mountain. Have you ever seen a more spectacular site in winter or in summer? Whether you trudge up the Grouse Grind or take the gondola, the view is truly impressive. Grouse Mountain in North Vancouver must be one of the few mountains in the world where you can take a city bus to the ski slopes.

At the top of Royal Oak at Oakland in Burnaby. In the early morning, particularly in spring, to the east one can see the sun peering through the mist over Deer Lake, to the north are the snow-capped mountains, and while it's hard to think of a cemetery as beautiful the park-like location down the road to the north certainly seems peaceful.

Capilano Suspension Bridge in North Vancouver. Can you say you have made it all the way across? Have you tried the new, narrow Cliff Walk or the Tree Tops Adventure? Pretty spectacular! "Voices from Vancouver's past mingle with the sounds of nature; beautiful gardens skirt colorful totem poles." The Lynn Canyon Suspension Bridge is pretty cool too, and it's free!

Steveston Village in Richmond. Working fishing boats, surrounded by charming shops and excellent restaurants, plus an aroma of the sea as you stroll along the boardwalk or walk out to Garry Point.

Stanley Park in Vancouver. How could I have not mentioned this already?! This is one of North America's largest and most impressive city parks. A thousand acres of playground with a myriad number of views and things to do. Walking, running or biking on the seawall is an exhilarating experience. Visit the Aquarium or feed the ducks in Lost Lagoon. If you have half an eternity, watch a cricket game. You get the picture—lots of pictures.

Bowen Island—is it a lifetime away, or just a 20 minute ferry ride from Horseshoe Bay in West Vancouver to Snug Cove (which must surely be two of the most picturesque ferry terminals in the world)? It's a great day trip, or stay longer. Or you can visit one of the other Gulf Islands.

Redwood Park. Yes, within the metro Vancouver area we have hundreds of massive beautiful redwood trees reaching to the sky. Have a picnic and then walk along the trails. Touch the soft bark of the redwood trees and see the more than 25 other kinds of trees in the park.

Victoria's Inner Harbor and the Empress Hotel. Whether you come for high tea at the Empress, just enjoy the stately atmosphere, or stroll along the harbor, you get a taste of a relaxing yet bustling part of beautiful BC.

Whistler and Blackcomb. Enjoy world class skiing or biking or world class relaxing with a taste of European charm in the village, or view the spectacular scenery from the Peak 2 Peak Gondola.

Barkerville. Billy Barker found gold here in 1862. At the peak of the Gold Rush Barkerville was one of the largest towns in the west. Today it's an historic and whimsical place to visit.

The Sea to Sky Highway from Vancouver to Whistler is spectacular but you see even more fabulous views on BC Rail and the Rocky Mountaineer. Howe Sound is the quintessential and splendid west coast fjord. Perhaps one day the Royal Hudson will again steam its way on the route as well.

Pacific Rim National Park and Tofino. The Pacific Rim is truly one of the most spectacular and rugged places in the world. Walk along the miles of beach. Camp near the beach or stay in a luxurious hotel. Enjoy the raw fury of nature during a ferocious storm. Enjoy the casual atmosphere of Tofino or go whale watching. Or hike the gruelling and majestic West Coast Trail.

Westminster Abbey. No, you don't need to go to London; just to Mission, BC. The Benedictine Monastery welcomes visitors and you can spend a few minutes in reflection and worship in a quiet atmosphere that reminds one of the glory of God.

Hell's Gate. Visit the Fraser Canyon and take a jet boat ride or a white water river raft on the Fraser as it plunges and cascades to the sea. Take a sedate tram ride and watch for the fish ladders symbolic of man's attempt to co-exist with nature.

Othello Tunnels. These tunnels are a few miles beyond Hope, BC, a reminder of a once thriving railroad, as well as a reminder of the huge amount of snow that can fall in this area. Watch the rushing, twisting river as it carves its way through solid rock.

Westminster Quay, Lonsdale Quay and the Granville Island Market. You may purchase fruits and vegetables and crafts and coffee, but as a bonus you receive atmosphere and relaxing views of the water.

Fort Langley. Built in 1839 by the Hudson Bay Company as a strategic Trading Post, the fort gives a fascinating picture of our pioneer history. Or visit the quixotic antique shops in town.

Hazelmere Campground in Surrey. Just a few minutes out of Vancouver you can be a world away in a secluded little campground. Wander through the meadows, play in the Little Campbell River, or explore the woods and at night enjoy a campfire.

Mt. Baker. Oh, I know that this magnificent perpetually snow-capped isn't in BC. But drive east on highway 1 just from west to east of Abbotsford or north on Whatcom Road or a dozen other places in Abbotsford to see the Sumas Prairie and how the mountain dominates the Fraser Valley landscape.

Have you seen the Filberg Festival in Comox or the Edge of the World Music Festival in Haida Gwaii or The Bulkley Valley Exhibition or the hot dog stands at the PNE? I could go on and on, but this isn't a BC travel essay. Enjoy Beautiful BC!

VIGNETTE 85

Adjectives, Adverbs and Beautiful Churches

MANY HUMAN BEINGS have an insatiable desire to acknowledge a power greater than themselves, often expressed as a desire to worship God. This can be done anytime, anywhere but often it seems we feel compelled to build a beautiful building which helps express the honor that we feel in approaching God and the respect that is due Him. The Old Testament understands this when it describes in inordinate detail how Solomon's temple was to be built.

We visited St. Peter's Basilica in Rome, the largest church in the world, and its exalted beauty and ambience certainly invites worship. St. Patrick's Cathedral in New York and Notre Dame Cathedral in Montreal are similar. The inside of these churches is shaped like a cross, and the arches are intended to reach symbolically up to the sky, and even all the way up to God.

We also visited St. Paul's Cathedral in London, and it is even more beautiful. We took a guided tour of the church, including a strenuous climb up to the dome. This was called "whisperer's wall" because the acoustics are so perfect that one can whisper at the wall and the sound of your voice curves all the way around, so that a person standing several hundred feet away around on the other side of the dome can hear you. This is a powerful metaphor showing that God can always hear us. We visited late on a Saturday afternoon so we were able to stay for Evensong. The service was plain and humbling, and yet it was exquisitely beautiful

Adjectives, Adverbs and Beautiful Churches

as the little lamps highlighted the choir as it melodiously led the worship. It was very inspiring and we could feel God's presence.

These famous cathedrals are impressively beautiful, but I have to say that the absolutely most beautiful church that I ever was in was the Bawlf Lutheran Church. We lived in a Norwegian community when I was growing up and my family attended this church in the small village of Bawlf southeast of Edmonton, Alberta. I'm sure that there is a certain amount of sentimentality in stating that this was the most beautiful church ever, but the feeling of God's presence was inviting and profound. The craftsmanship was extraordinary, with various carvings and impressive gold paint trimmings. The inside of the church was a rectangle, but the main aisle went kitty-corner from the main entrance doors in one corner to the altar and platform at the opposite corner. This meant that the first rows at the back were very short, seating only two or three adults, and the rows gradually got wider until near the front they again started to get shorter. The distinctive European influence was demonstrated by the choir loft that was elevated along one side wall. Along the other side wall was a very long single bench. I can still remember sitting on this bench one Christmas Sunday morning marvelling at the grandeur and the sensational beauty of this church, and instinctively knowing deep within me that God was even more glorious and awesome and as we recalled His unique sacrificial Christmas gift to us, and that He was pleased with our worship that day. It was in this church that I became a Christian at age 11. It was a tragedy of unspeakable proportions when this church burned down one winter night a couple of years later. They built a new church, and God was in it also, but after the grand architecture of the magnificent old church the new one seemed like a plain and drab warehouse to me. It was like comparing the Mona Lisa to a mundane kindergarten painting.

Another extraordinarily beautiful "church" in which I worshipped was at the top of Sulphur Mountain. When I was at the University of Alberta in 1964 our youth group went on a week end trip to Banff. On the Sunday morning we took the chairlift to the top of Sulphur Mountain and both literally and spiritually we had an awe-inspiring mountain-top worship experience. This reminded me of Jacob's desire, as expressed in the Bible, to build an altar after his dream about a stairway to heaven (if he had lived in the 1960's he would have dreamt of a chairlift reaching all the way up to heaven) and it illustrates the human desire to reach up to heaven and to build monuments to recognize God's presence. Jacob called his place "Bethel" meaning house of God because he said, "Surely the Lord is in this place." We said the same thing about Sulphur Mountain that day.

I would like to tell you about two other beautiful churches. I was in Los Angeles to teach a laboratory quality assurance course some years ago, and while I was there I had arranged to check out a ministry to inner city people since our kids' church youth group was planning to be trained and visit there for a week. A young couple picked me up at the airport on a Sunday morning and took me to this "church" which was "just" a high school auditorium. There were about 2500 people there and I counted five white folks including the three of us. The worship service was active, but I felt totally comfortable; people were simply praising God. The Pastor got three standing ovations, and this was just his opening prayer! The black choir of about 80 men and women was phenomenal and absolutely awesome—they could really sing! I have trouble singing songs that have more than one note but even I could visualize singing like this happening in heaven. We sang praises and worship songs for about an hour, and then the Pastor preached for more than an hour. At least seven times the congregation stood and enthusiastically applauded and said, "Amen," or, "Preach it, brother." I kept smiling both inside and out because I knew that God was in this place and in these people. It was an unforgettable experience and it was almost breathtaking both to see God being honored and at work. And communion, with 25 deacons serving, was immeasurably meaningful and beautiful. That church was beautiful because the people were beautiful and their worship was sincere.

The last remarkable church that I want to mention is our present church—Northview Community Church in Abbotsford, BC. This church doesn't have extraordinary architecture and it isn't hundreds of years old and it isn't shaped like a cross, but is does have awesome pastors and sincere people who have a deep desire to know and serve and honor God. I have come to the conclusion that a beautiful church is anywhere that people gather to fervently worship God.

VIGNETTE 86

Quebec Wants a Divorce with Bedroom Privileges

IN 1995 THE province of Quebec held a referendum asking their voters whether Quebec should secede from Canada and become an independent state. The referendum question basically asked if Quebec should secede and become a sovereign nation and ask (demand?) a new economic and political partnership with Canada. The motion was defeated by a very narrow margin of 49.4 percent "Yes" to 50.6 percent "No". Before that, an unsuccessful 1980 referendum question proposed to negotiate "sovereignty-association" with the Canadian government, while the 1995 question proposed "sovereignty", along with an optional partnership offer to the rest of Canada.

After Quebec almost answered yes to separation, other Canadians and the government realized that they should take some initiative, since continually appeasing Quebec wasn't working. The federal government drafted "The Clarity Act" that established the conditions under which the Government of Canada would enter into negotiations that might lead to secession following such a vote by one of the provinces. Before passing this Act the government asked the Supreme Court whether any province had the right under the Canadian Constitution to declare independence. The Supreme Court said yes, but that a unilateral declaration of independence was illegal. The terms of the Clarity Act makes it more difficult for a province to separate from the Canadian federation.

Canada repatriated and amended in 1982 the BNA Act (British North America Act) of 1867 that established Canada as a sovereign nation and changed its name to the Constitution Act, 1867 (the new Act also included the Canadian Charter of Rights and Freedoms). Quebec was the only province that refused to approve this new Act. In 1987, then Prime Minister Mulroney attempted to gain Quebec's acceptance of the 1982 Constitution Act by negotiating the Meech Lake Accord with the provinces, which recognized Quebec as a "distinct society" and obligated the federal government to consider Quebec's unique status, plus giving Quebec a constitutional veto in federal matters. However the government failed to obtain the required unanimous support of provinces. They were concerned about some of the special powers that would be given to Quebec and the deal fell apart. Then in 1992 the government negotiated the Charlottetown Accord, which proposed the transfer of some powers to the provinces from the federal government and greater taxing authority for them, as well as a new constitutional arrangement with Quebec. But this was rejected in a national referendum because Canadians were still not comfortable with the transfer of powers that would have been given to the provinces, and especially the new constitutional arrangement with Quebec. The latest chapter in this series occurred in 2006. Prime Minister Harper asked Parliament to pass a motion declaring that Quebec was a "nation" within a united Canada.

Where does all this leave us, and why do some Quebecers still want to separate from Canada? Some Quebecers feel they need to protect their language, culture and unique legal system, since they believe that Canada cannot properly do this. They say that Canadians outside Québec don't fully appreciate that Quebecers are different and have a somewhat isolated culture that makes them "separate and distinct." They are defensive and insular since they fear that their culture will be overrun by both Canada and the U.S. Even France, while sympathetic to their situation, looks down at them. Canadians outside Quebec see them as complainers who want to get more for themselves and who don't want to contribute to Canada.

Even though the "Quebec situation" is now relatively stable, there still are many questions. Has there been sufficient contingency planning in the event that Quebec would separate? For example, would they print their own money, have their own banking system, and establish their own armed forces, etc.? The 2011 problems with the Euro illustrate the hazards that can occur when sovereign nations share a currency but not a common fiscal or monetary policy. How could Quebec expect to maintain ties to Canada for these issues? That is, does Quebec want a divorce, but still expect "bedroom privileges?" What jilted lover has ever granted such a selfish and arrogant request? Would large crown corporations, such as Air Canada and Canadian National move out of Montreal? How would Canada be compensated for all of the federal buildings that are now in Quebec? If Quebec separates, would further fragmentation of Canada occur? What would happen to the Maritime Provinces? What would the Western Provinces do, since there has often being a feeling of alienation there as well? What does the U.S. think about the possibility of a divided and unstable Canada?

The main argument against separation is economic. In spite of its hydroelectric resources, some industrial strength and significant mineral wealth, Quebec remains a 'have not' province. While Quebecers pay considerable taxes to Canada, Ottawa already invests more than that in the Québec economy through a huge range of projects. Many Quebecers fear that separating from Canada could mean a decline in their standard of living. Federalists argue that Quebec's special status can be best protected within a strong Canadian federation. They say that Quebec will be worse off if they are independent and lose the protections that the constitution provides. "Canadians" could choose not to buy Quebec products as a way of protesting separation. The deeply emotional need for Quebecers to be independent is balanced out by the realities of its economic ties to Canada and opposition within Quebec to the idea. The consensus in 2011 seemed to be that Quebec will not separate, since the financial stakes are too high for Quebecers to take that plunge. It seems more likely that Quebec will continue to strive to obtain more autonomy, rights and constitutional privileges to protect their culture and heritage. That will allow it a kind of government different from the other provinces and territories, even though this will cause resentment in the rest of Canada. It seems like we will be discussing this problem for a long time yet.

VIGNETTE 87

"Check Your Value System at the Door"

SOME YEARS AGO, Jeffrey Simpson of the Toronto Globe and Mail felt it necessary to declare in his column that Stockwell Day, then recently elected as a Reform Party MP from Alberta, should "check his [Christian] value system at the door" [of Canada's House of Commons]. Why was it that MacLean's magazine also felt justified is asking. "How scary is Stockwell Day"? Did anyone ever ask Pierre Elliot Trudeau to check his "liberal" secular humanist values at the door? Did anyone ever ask MPs Jack Layton or Ujjal Dosanjh to check their personal socialist values at the door?

Why is it that the generally liberal media feel justified and immune from criticism when they encounter and disparage individuals with Christian values? Why is it that Christian values are often challenged, and why is it that secularists seem threatened by those who have Christian value systems? Being a social conservative today risks incurring the wrath of the secular elite. The mantra of those who would impose their liberal views on society is that only their views are valid and that conservative values are suspect.

There are several possible reasons why liberals are suspicious of social conservatives. A significant factor, I believe, is human pride and subconscious guilt. No one likes to be set against a high standard, and no one likes to acknowledge that they have done wrong things, or had morally wrong thoughts. A natural reaction is to lash out at those who

"Check Your Value System at the Door"

support strong fundamental values such as Christian beliefs. At the same time, liberal secularists seem to only challenge Christians but not other belief systems.

Many people who say they are tolerant turn out to be intolerant of Christian values, and at other times what they really mean is that they are tolerant of everyone who agrees with them. In our pluralistic and multicultural society it's necessary to recognize that many Canadians have different points of view on social issues, and we need to respect everyone.

Another significant factor, unfortunately, is the lack of tolerance and grace sometimes demonstrated by many Christians, who really should know better. Phillip Yancey has written a thought-provoking book called *What's So Amazing About Grace?* in which he chronicles examples of where Christians have hurt the cause of Christianity by not showing grace, by not showing tolerance of others (which is not to be confused with acceptance).

What can the Christian community do about the bias that liberals have regarding social conservatives? Well, a start is to respect others while standing up for one's own beliefs. Show grace to those who criticize us. Practicing grace will do much to improve society's perception of Christians and Christianity. Another important thing is for Christians to be active in our communities to show that we care—at city council, on the school board, in charitable organizations, at the food bank and so on. Be informed about public issues and form a reasoned Christian response. Write reasoned letters and e-mails to the editor and to columnists (many columnists are open to positive and informed opinions even if given from a differing perspective). Write graciously and write reasonably often.

In conclusion, it's OK to challenge misinformation and incorrect perceptions of Christianity, or anyone else with whom you disagree, but do so in an informed and gracious manner. Grace is one thing that secular society finds difficult to duplicate.

VIGNETTE 88

Tantalum

YOU HAVE PROBABLY never heard of tantalum, but chances are high that it's part of your everyday life. I'm not sure where I first heard about it (I'm not really a science geek), but the name intrigued me so I checked Wikipedia for some information. Tantalum is a chemical element with the symbol Ta and atomic number 73. Tantalum is a rare, hard, blue-gray, lustrous transition metal that is highly corrosion resistant and is widely used as minor component in alloys. The chemical inertness of tantalum makes it a valuable substance for laboratory equipment and a substitute for platinum, but its main use today is in tantalum capacitors in electronic equipment such as mobile phones, DVD players, video game systems and computers.

The name tantalum was derived from the name of the mythological Tantalus, the father of Niobe in Greek mythology. In the story, he had been punished after death by being condemned to stand knee-deep in clear water with perfect fruit growing above his head, both of which eternally *tantalized* him. (If he bent to drink the water, it drained below the level he could reach, and if he reached for the fruit, the branches moved out of his grasp.) The discoverer of this metal, Dr. Anders Ekeberg, called his discovery tantalum because of its incapacity, even when immersed in acid, to absorb anything and become saturated.

Tantalum is dense, ductile, very hard, easily fabricated, and highly conductive of heat and electricity. The metal is renowned for its resistance to corrosion by acids; in fact, at temperatures below 150° C tantalum is almost completely immune to attack by the normally aggressive aqua regia (one part nitric acid and three parts hydrochloric acid). However, it can be dissolved with potent hydrofluoric acid or a few other acidic solutions. Tantalum has a very high melting point (3017 °C).

The primary mining of tantalum is in Australia and other countries such as China, Ethiopia, and Mozambique, which mine ores with a relatively high percentage of tantalum. Tantalum is also produced in Thailand and Malaysia as a by-product of tin mining. The major use for tantalum, as the metal powder, is in the production of electronic components, mainly capacitors and some high-power resistors. Tantalum electrolytic capacitors exploit the tendency of tantalum to form a very thin protective oxide surface layer, using tantalum powder, pressed into a pellet shape, as one "plate" of the capacitor, the oxide as the dielectric, and an electrolytic solution or conductive solid as the other "plate". Because the dielectric layer can be very thin, a high capacitance can be achieved in a small volume. Because of the size and weight advantages, tantalum capacitors are attractive for portable telephones, personal computers, and automotive electronics.

Tantalum is also used to produce a variety of alloys that have high melting points, are strong, and have good ductility. Alloyed with other metals, it is also used in making carbide tools for metalworking equipment and in the production of super alloys for jet engine components, chemical process equipment, nuclear reactors, and missile parts. Because of its ductility, tantalum can be drawn into fine wires or filaments, which are used for evaporating metals such as aluminum. Since it resists attack by body fluids and is nonirritating, tantalum is widely used in making surgical instruments and implants. For example, porous tantalum coatings are used in the construction of orthopedic implants due to tantalum's ability to form a direct bond to hard tissue.

There is so much information in our world. It's fascinating to me to realize how little we really know and are yet able to live normal, productive lives.

VIGNETTE 89

No One Likes Food Additives, but Everyone Wants to Use Them

I USED TO work for Health Canada and periodically gave a few lectures on food additives, so I have an interest in and watch the news for reports on this topic. Almost no one, when asked in a survey, says that they wanted to use food additives. But yet basically every one of us use them every day—color in orange peels (we want oranges to be consistently orange, not mottled orangey-green), preservatives in bread (we don't want our bread to go mouldy in two days), firming agents in pickles (we like the crunchiness), bleaching agents in flour (we want our flour to be white) and aspartame or acesulfame K in our soft drinks (we need to reduce our calorie and sugar consumption). We don't want additives, but we want or insist on using products that contain them.

Food preparation in the home has changed dramatically in the past 100 years in North America. People don't have chickens in their backyard, they don't have a couple of cows, and most people no longer have a garden. When my mother or my grandma prepared dinner, it took pretty much all day. Somebody would kill a chicken, and if it was summer one of the kids would pick some beans or carrots or corn, and lettuce, and dig some potatoes from the garden, and by 6 o'clock dinner would be ready. (If it wasn't summer, the beans or corn or peas etc. would have been canned in fall, lettuce wasn't generally available, and the meat was cured or frozen). Our typical grocery supermarket today may contain as many as 10,000 different products, while the general store a hundred years ago had perhaps a hundred different items.

No One Likes Food Additives, but Everyone Wants to Use Them

Today Mom and Dad come home at 5.30 and are eating dinner by 6.30. Much of our food is processed in New Jersey or Romania; or it is grown in California or Chile or Ecuador or Mexico and must be transported to our grocery shelves 12 months a year, appearing fresh and attractive (consistent in appearance). We expect variety and quality and an assurance of food safety. In other words, most people want the advantages that food additives provide, but if you ask them they will say that they don't want any food additives. People are suspicious of food additives, and this is OK, but they also want the advantages that additives provide. Preservatives increase shelf life and retard mould growth, other additives ensure a uniform texture and appearance, while others make it possible to have aerosol products as in "PAM" or whipped cream.

Grandma didn't necessarily know that she was using food additives or chemicals. She thought she was just preparing dinner using conventional ingredients. She just added egg yolk to vegetable oil and vinegar to make some "salad dressing", or she added sodium bicarbonate (baking soda) to raise the dough when she baked bread. Or she added some rice grains to the salt shaker or container to prevent clumping, and she used pectin to make a firm jam or jelly from the berries grown in her garden. Today we have purified additives (pure chemicals added to the product) that do all of these jobs—our salad dressings have emulsifiers, the salt has an anti-caking agent, the jam already contains pectin or a thickening agent, and so on. What distinguishes food additives from conventional food ingredients used in the "old days" is that the additive now is used as a relatively pure chemical.

What, if anything, should be done? Most of us need to be autodidactic (learn about a subject without the benefit of a teacher or formal training) with regards to food additives. We need to find out what procedures scientists and government food regulators use to approve, or reject, food manufacturers' requests to use a given additive in their food product. Food additives are assessed in much the same way as drugs to determine if they are safe to use. When a food processor submits a request to use an additive, they need to provide animal toxicology data, identify all the foods in which they want to use the additive, and calculate how much of the additive a person would consume per day with a standard diet. If this "expected daily intake" exceeds the "tolerable daily intake" determined by toxicology studies, then the additive would not be approved.

Just a few years ago some folks proposed something they called the 100-Mile Diet. This became fairly popular, inspiring many people to change the way they eat. Purchasing locally raised and produced food should be fresh, better for the environment since long distance transportation isn't required, and it should benefit local farmers and local economies. If we are serious about this concept then we have to make sure that we preserve our local farms and market gardens. Purchasing "organic food" is another option, but I don't believe that this is really necessary. Contrary to some of the "propaganda" out there, food is not "laced with dangerous chemicals" such as pesticides because of the regulation of food additives and contaminants. Reducing our expectations might be a good idea too—how is it that we expect food transported halfway around the world to be as fresh as if it were picked yesterday, but not contain any preservatives or anti-ripening agents?

Eating healthy does take a bit of work, but it is possible even using food additives. Limiting our portions seems to be even more difficult for a large segment of the population!

VIGNETTE 90

MRSA—The Four Scariest Letters in the Alphabet

WE PERIODICALLY HEAR reports about people getting seriously ill while in the hospital. Sometimes this is due to Methicillin-resistant Staphylococcus aureus (MRSA) infection, which is caused by any strain of staphylococcus bacteria that's become resistant to the commonly used antibiotics, including various penicillins and cephalosporins used to treat ordinary staph infections.

Methicillin is a very potent antibiotic used only when absolutely necessary, and bacteria that are resistant to methicillin are called "superbugs". One major reason that superbugs develop is the over-prescription of antibiotics, and there are two usual suspects responsible for this—doctors and patients. When physicians prescribe antibiotics for unnecessary conditions (such as viral infections or an ordinary cold), it actually promotes the natural mutation of bacteria, ultimately resulting in the creation of new, resistant strains that can often be easily passed from one patient to another in a hospital setting. But patients also bear a major responsibility, since they often demand a prescription for almost any malady and feel short-changed if the doctor just tells them to go home and rest and drink lots of fluids. We need to be more proactive as individuals, and as society in general, in strongly questioning the over-prescription of antibiotics. Perhaps medical schools and teaching hospitals also need to be more diligent in training new doctors how and when to prescribe antibiotics, since it seems that more and more people are being infected with superbugs and are killed by them.

Most MRSA infections occur in people who have been in hospitals or other health care settings, such as nursing homes and dialysis centers, especially since these patients may have a weakened immune system. MRSA infections typically are associated with invasive procedures or devices, such as surgeries, intravenous tubing or artificial joints. These people might actually die from a non-superbug infectious disease by a secondary infection, but if they are struck with a superbug, their odds of surviving the infection drop even further. Another type of MRSA infection, which is spread by skin to skin contact, has occurred in the wider community among healthy people. Other at-risk populations include groups such as high school wrestlers, child care workers, and people who live in crowded conditions.

Hospital infections are a concern. I read that the World Health Organization (WHO) reported that hospitals in Canada have some of the highest infection rates in developed countries. The rate in Canada was 11.6 percent; while it was only 4.5 percent in the U.S. and 7.1 percent in Europe (Germany at 3.6 percent was the lowest country in Europe). There have also been other reports that doctors in hospitals need to wash their hands more often. There are other components to providing safe and, where necessary, sterile environments in hospitals, but surely we can do better and wash our hands.

It's easier said than done, but the best thing to do is protect your own health so that you can stay out of hospitals if at all possible. There are a lot of sick people in hospitals (!) so it's best to stay out of them if possible, and the best way to avoid hospitals is to prevent disease and obesity by taking responsibility for your own health. Practicing good nutrition, getting lots of exercise and perhaps taking vitamins will help.

Staph. aureus most commonly colonizes the nostrils, but the rest of the respiratory tract, open wounds, intravenous catheters, and the urinary tract are also potential sites for infection. Healthy individuals may carry MRSA without any symptoms for long periods of time. In most patients, MRSA can be detected by swabbing the nostrils and isolating the bacteria found inside. Combined with extra sanitary measures for those in contact with infected patients, screening patients admitted to hospitals has been found to be effective in minimizing the spread of MRSA in hospitals.

MRSA—The Four Scariest Letters in the Alphabet

MRSA may progress substantially within 24 to 48 hours of initial topical symptoms. After 72 hours MRSA can take hold in human tissues and eventually become resistant to treatment. The initial presentation of MRSA is small red bumps resembling pimples, spider bites, or boils that may be accompanied by fever and occasionally rashes. Within a few days the bumps become larger, more painful, and eventually open into deep, pus-filled boils. But in some cases vital organs can be affected and lead to widespread infection (sepsis), toxic shock syndrome and necrotizing ("flesh-eating") disease, and in other cases damaging the valves of the heart. Doctors don't yet know why some healthy people develop MRSA skin infections that are treatable, whereas others infected with the same strain develop severe infections or die.

There has been considerable effort in hospitals for doctors and nurses to wash their hands and to use surface sanitizers. This of course should be practiced by the public as well, not just in hospitals but in public washrooms and other places.

VIGNETTE 91

Reforming the Senate

TO REFORM OR abolish or do nothing—that's the question facing our politicians about the Canadian Senate. Prime Minister Harper, once he obtained a majority in the House of Commons in 2011, indicated that he wanted to move forward in an attempt to reform the Senate, our favourite whipping body of near-octogenarians. This is a gutsy move for several reasons. The more conventional tactic would be to change our Constitution, but given past failures and the pretty clear guarantee that at least some of the provinces would derail the process, Harper chose the back door approach.

Some years ago the Reform Party had advocated a "Triple E" Senate (elected, equal, effective), but would you believe that 0 out of 3 is now more likely than ever? One out of the three (elected) is a possibility; "effective" is up for debate even if Harper's reforms are accomplished, but "equal" is even less likely. The Atlantic Provinces entered Confederation with a disproportionate number of seats (e.g. PEI with a population of 143,000 had four Senate seats in 2011 and New Brunswick with a population of 750,000 had six seats, while BC with 4.5M citizens had only six seats, and none of these Eastern provinces are remotely likely to give up this advantage. Changing the representation formula would require a constitutional amendment, which as we have seen is not going to happen any time soon.

Mr. Harper's proposed legislation had two main features, namely elections and term limits of possibly nine years. In the 1960s then Prime Minister Lester Pearson capped the maximum age of Senators at 75, so establishing a maximum term length shouldn't be a giant step. The bill that was drafted proposed a voluntary framework by which provinces would hold Senate elections. Technically and procedurally, the Governor-General "on the "advice of the Prime Minister" still would have to make the actual appointments, but at least if the election process were used he would make the appointments from the winners from each province. But Quebec would likely challenge the bill in court (surprise!) while some other provinces simply want to abolish the Senate (and even this option requires the unanimous consent of the provinces, and where the provinces are concerned "unanimity" is a foreign concept except when asking for more money or more power). As of 2011 there were actually at least two "elected" Senators who were appointed in the usual fashion after winning special elections in Alberta. Some Opposition politicians grumble that having some elected and some appointed Senators would be a mish-mash problem if some provinces refused to have Senator elections, but at least having some elected Senators would be a start, particularly if they would be high calibre individuals who worked hard and were effective.

Why reform the Senate? Well, a reformed Senate may gain more respect from Canadians, and it could dispel the anonymity, perceived arrogance and lack of accountability of appointed Senators. An elected Senate could actually become more credible and more powerful as a "chamber of sober second thought." Presumably legislation would also be enacted to give the House of Commons the necessary power so that when differences occur the U.S.-style upper house/lower house stalemates would not be possible. An elected Senate would also go some distance in satisfying the populist feeling, in the west at least, that Senators, like other politicians, should be accountable and that there should be at least some semblance of regional representation. Abolishing the Senate, while having some tax saving appeal, probably isn't the right thing for Canada. Canada is a huge country geographically, with the population concentrated in two provinces, so a counter-balance body to the House of Commons where there is representation by population may still be a good idea.

What does the Senate do? The Senate reviews bills passed by the House of Commons, and passes, rejects or suggests amendments to those bills. The Senate also proposes its own legislation, though it cannot propose bills that would appropriate revenue (as money bills are the exclusive right of the House of Commons). No bill passed by the House of Commons can become law without the agreement of the Senate, with the exception of some types of constitutional amendments where the Senate has a temporary 180-day veto only. Senators are also known for conducting comprehensive and thorough studies that are valued and used by organizations throughout Canada, including the elected House of Commons. The Senate actually does some useful Committee work in studying and reporting on issues of concern to Canadians, so its work in the long term may actually benefit Canadians. The Senate is responsible for protecting regional interests (as its seats are distributed by region, rather than by proportional representation), and for protecting minority interests. As far as I understand it, the purpose of a second legislative body like the Senate is to serve as a check and balance in our government system. It seems that we will be "stuck" with the Senate for a long time, regardless of what if any changes are made, so we need to hope that it does a good job and is worth the money that it costs.

VIGNETTE 92

Plastic Money

NO, NOT CREDIT cards. We're talking real money here, as in hundred dollar bills and $50s and $20s. The Bank of Canada has a program to shift to the production of synthetic plastic banknotes, starting with the introduction of $100 bills in November 2011, followed by $50 bills in March 2012 and $20 bills later in 2012. There are at least four primary reasons for the change from the current cotton-based banknotes. The first is that the Bank of Canada and the RCMP hope that the use of the polymer based plastic bills will be substantially more difficult to counterfeit. A second major reason is that the plastic money is two to four times as durable, resulting in both reduced production costs and environmental benefits in reducing energy consumption by running the printing presses less often. A third benefit is that these plastic bills don't curl or fray at the ends, thereby causing fewer jams in ATMs and automatic bill counting machines. A further advantage is that these bills can be recycled into other plastic products when their lifespan is finally reached.

You may think that the move to more durable banknotes is a bit too late given our not so gradual move to a cashless society, but the bean counters in Ottawa have calculated that half of all financial transactions still involve cash. At least 30 countries have already made the move to plastic bills, or are in the process of doing so. While the Bank of Canada issues all Canadian bank notes, the actual production is contracted to two companies based in Ottawa.

The printing process for plastic bills is quite interesting. A thin polymer is created from plastic chips, and these rolls go through a press that applies multiple layers of white ink to each side, producing an adhesive surface. Small transparent spaces are inserted to dot these sheets, and a laser burns a hologram-like image onto each space, which makes the bills practically impossible to scan or photocopy. Background color is applied; additional layers of ink are applied to create details like the outline of a portrait and then finer details of the design. Varying the thickness of the ink will give the bill a ridged texture which can be detected by hand, while also making counterfeiting more difficult.

The advantages detailed above are all important, but the driving force has been the need to make counterfeiting of bank notes much more difficult. Increasing security features on existing bills and training retailers to identify fake bills has already reduced the number of counterfeit bills that are circulated each year, but this number is still too high, and much higher than in countries such as Australia which introduced plastic bills over 20 years ago. Counterfeiters, like many criminals everywhere, are persistent, creative and imaginative (using laser printers and advanced printing techniques, for example), but officials are confident that this new approach will make counterfeiting significantly more difficult. Many retailers have refused to accept our conventional $100 bills and even $50 bills, and when they do so the customer often has to first stand and nervously watch while the clerk carefully checks to see if he or she is perceived as a scam artist passing off a bogus bill.

Once the new plastic bills are in circulation, the Bank of Canada will work with regular banks and other financial institutions to remove the old bills as quickly as possible, although the regular bills will always be accepted as long as they are in good shape. Perhaps the central bank could also work on making any of our banknotes valuable enough that we can actually buy something with them. Or perhaps they will soon work on Canadians to give up the penny.

VIGNETTE 93

Will that Be Credit or Overdraft?

NO ONE LIKES taxes, but the vehement antipathy of many citizens towards taxes is causing major problems around the world to politicians who want to implement necessary legislation to pay for services but also want to get re-elected. Unfortunately, the politicians' desire for power often overwhelms the desire to do what's right, and municipal, provincial (or state) and federal governments find themselves overspending and therefore with huge deficits. The unwillingness or inability of governments to reduce spending and collect the taxes that are required to maintain programs and pay down the debt creates huge problems for those countries. Nowhere was this more evident that in the U.S. and Greece. And we heard a new acronym in 2011, namely "PIGS" which stood for Portugal, Ireland, Greece and Spain, all countries where their proclivity to spend exceeded their capacity to pay their bills. I guess they didn't figure a way to add 'U.S.' to the acronym.

Diane Francis of the National Post reminded us in June 2011 that the U.S. is the country that had the famous Boston Tea Party; throwing tea into the harbour and launching a revolution over the issue of taxes. Ms. Francis indicated that after the War of Independence in the U.S. the 13 state governments at the time were unable to get anyone to pay taxes, so they raised the needed operating money by taking land from the Indians and selling it to the flocks of new settlers. Eventually some taxes were of course initiated and paid, but Americans are "still loopy about taxation." It is said that if Americans paid the same taxes on incomes, gasoline, alcohol and cigarettes as Canadians pay, they wouldn't have a deficit problem. There are many situations where someone earning $250,000 a year or more pays fewer taxes, at least on a percentage basis, than low income single parent families.

Greece has an even more serious problem. Taxes were considered by the Turks (who ruled Greece at one time) as a symbol of subjugation, thereby building strong resistance among the Greeks to taxes. Then after a series of Greek monarchs and dictatorships didn't help the situation, Greece became a Republic in 1975 but the successive governments seemed unwilling or unable to collect the taxes needed to run the country. As a result Greece has a massive deficit and has needed the European Community to bail them out to avoid defaulting on their huge loans. Now that the EC is balking at doing this again and again, Greece may actually go bankrupt. And what's the reaction of many of the Greek citizens? Why, there's rioting in the streets protesting government attempts at cutbacks to balance the budget by cutting wages and entitlements, as well as attempts to prosecute tax evaders (a new concept!) since the people refuse to give up their entitlements or actually pay for them!

Over the past 30 years we have seen the development of a "culture of entitlement" in many countries. This was particularly evident in Greece in the spring of 2011 where the combination of overly generous social benefits and insufficient tax revenues brought the country to the verge of bankruptcy. Italy has had some serious problems as well where many people retired at age 50 and now are collecting pensions with a reducing number of working people, plus there apparently are many non productive municipal jobs in towns and villages that need to be supported by tax payers. These problems aren't limited to Greece and the other EU countries in financial difficulty. Around the world, workers are finding that, if they want continued salary increases while maintaining benefits, the result quite often will be lost jobs.

Canada also has its share of tax avoiders or tax cheats. Statistics Canada reported that there was $444 billion in unreported income during the 16 year period between 1992 and 2008. The taxes that should have been paid should have resulted in the rest of us paying lower taxes and/or seeing a reduction in the national debt. We often decry those who obviously evade taxes, but there apparently are a lot of ordinary Canadians who do this by supporting the underground economy.

The U.S. government also needs to face the music. In August of 2011 the government approved a watered-down deficit reduction plan but still voted to extend the existing debt limit to avoid being in default of loans and unable to pay for government services. The government was faced with three bad-to-worse choices or scenarios: a serious down payment plan that would require deferring programs, postponing the deadline (which they have done before but would make the situation even worse in the future), or doing nothing. Any of these options could lead to bankruptcy or at least "unchartered territory" that would have major economic consequences for the U.S. and the rest of the world.

Neil Reynolds of the Toronto Globe and Mail reported in June 2011 that the U.S. federal government spent $30,000 per household and State governments spent $25,000 for a total of $55,000 per household in 2010. The problem was that the federal government borrowed $12,600 per household and without any borrowing these governments would have only been able to spend $17,400 per household. This sounds like it could be a problem! People expect $5 in government services for $4 worth of taxes as a fundamental right.

Canadians have much lower federal government debt (approximately $274 billion, which is still a huge number!); one estimate is that this translates to $22,000 per household, of which only $2,500 was borrowed. The Canadian government has a plan to balance their budget within several years, while the U.S. is actually considering raising its debt limit. Canadian Provinces spent $24,800 of which $1,600 was borrowed. So the U.S. spent just over $9,000 more per household than Canada spent ($55,000 vs. $46,800). But Canadians do have one bigger problem—we have higher personal debt (148 percent of after-tax income, compared to 114 percent in the U.S.) so we do need to reduce our debt also to be better prepared for a financial downturn.

Stay tuned for further developments! Hopefully, Canada's approach to addressing its budget and deficit problems will be successful, but our problems pale in comparison to those of several European nations and the U.S. Plus we could be overwhelmed by the tsunami of economic consequences if things rage out of control in these countries.

VIGNETTE 94

Dirty Jobs and Hazardous Jobs

"IT'S A DIRTY Job but Somebody Has to Do It." No, we're not talking about lawyers or politicians here. Mike Rowe has made a career on TV of talking about, and trying out, dirty jobs. Some of his "dirty jobs" are just plain dirty and perhaps not too interesting, but these are jobs that honest people do to make a living, and not coincidentally, make life easier for the rest of us. A few of Mike Rowe's episodes or jobs include road kill cleaner, pig farmer, sludge cleaner, ostrich farmer, mushroom farmer, garbage collector, coal miner, asphalt paver, and big animal veterinarian. A few other jobs that he tried, such as a chimney sweeper or a termite researcher or a leather tanner, also don't sound too inviting.

Some of the dirty jobs he has looked at just cry out for more information. How would you like to be an alligator egg collector? Alligator farmers, who sell alligator hides, find that their alligators don't lay enough eggs to replenish their stock, so they need to hire alligator egg collectors who go into the Florida Everglades with their fancy flat bottom boats and then wade around to search for the nests to collect eggs in the wild. They say that alligators are good mothers (I'm not making this up), so it's a bit dangerous, shall we say. Or perhaps you would like to be a Las Vegas Casino food recycler? Before you stretch your imagination too much, let's just say it really is a dirty job and involves a pig farm. Then there is being a maggot farmer; they grow maggots to use as fish bait. The smell is, you might guess, interesting and it turns out that fishermen want pink maggots (perhaps they are catching female fish that like pink). Then there is the job of operating a "Spider Pharm" which involves collecting spiders and milking them for their venom. Who knew that spiders could be milked? And I suspect that you wouldn't want to be a leech trapper, collecting those cute little blood sucking worms that are also used as fish bait!

It turns out that only female chickens, called hens, are the ones that lay eggs (roosters are the guys that used to work as alarm clocks and get people up in the morning), so the job of being a chick sexer is important, but messy. With some operations hatching something like 80,000 chicks in a day, separating the boys from the girls early on is important. Without spilling the whole story we can tell you that chick sexing involves squeezing something out of each chick and assessing the nature of the excrement. It's too bad that they can't do like we do with people and wait until various behavioral and anatomical features start to manifest themselves.

I can relate to the dirty job of being a dairy cow midwife; when I was a kid I can remember helping my Dad deliver baby calves when the mother cow was having difficulty, and things got a bit messy. The initiation of offspring in dairy cows doesn't start with a live bull anymore; it often starts with an "artificial insemination technician" using very long gloves to insert bull semen into the appropriate place.

Working as a "hair fairy" doesn't sound very glamorous; this involves removing lice from human hair. One of the worst dirty jobs has to be that of a "septic tank technician" which not only involves collecting stuff from septic tanks, but also going inside the honey wagon to clean it. And I bet that no kid ever said, "When I grow up I want to be a worm dung farmer." Some folks actually operate "worm ranches" to collect worm poo, which is used as organic fertilizer and plant food; apparently these worms are easy to take care of since they are just fed animal manure and then stale bread and shredded paper for dessert. Nobody could make this up. Everyone needs to find their niche in life!

Leaving Mike Rowe aside, special kinds of dirty jobs are "hazardous jobs." According to 'CNNMoney', the ten most hazardous jobs are timber workers at 118 deaths per 100,000 workers, fishers (71 fatalities per 100,000), followed by pilots and navigators (we should hasten to add, not commercial pilots but primarily bush pilots, water taxis and crop dusters), driver sales workers (like pizza deliverers and vending machine fillers), roofers, electrical power installers, farmers, construction laborers, and finally truck drivers at 25 fatalities per 100,000 workers (truckers accounted for the most deaths but there are so many that the rate per 100,000 is lower). Jobs such as underwater welding, skyscraper window washers, crane operators, policemen and firemen all sound like some dangers are routinely involved. High school or middle school teachers haven't made any of the lists that I saw, but you have to be a brave person to have these jobs. So if you're sitting behind a desk crunching numbers or serving customers and think that your job is boring, take solace from the fact that you're not likely to be injured or killed when performing your duties. Just don't start playing computer games or using Facebook all day or you could experience another type of hazard—getting fired.

VIGNETTE 95

Painting Pictures of the Big Apple

WE WERE PLANNING to attend a conference in Philadelphia a few years ago and thought it would be a good opportunity to take a side trip to New York City. Two or three nights would be appropriate we thought, until we checked the prices of hotels in Manhattan. I called one hotel, a solid 2-star built in 1905, and the going rate was $323 US. We were almost resigned to not taking this little side excursion, but I finally found a Holiday Inn that charged only an arm and a leg, for 1 night. We made an adjustment in that we stayed in Philadelphia the first night and took the 6:46 am train to New York (going through some interesting places like Princeton, New Jersey along the way) arriving at Penn Station about 8:15. Permit me to paint some pictures of the Big Apple.

We trudged up the 40 mile long escalator with our two big suitcases and a briefcase up to ground level to find ourselves standing in front of the historic Madison Square Garden. The two wide-eyed Canadian tourists were instantly plunged into chaos and history. We had heard about hailing taxis, so I valiantly started waving since there seemed to be a lot of bright yellow cabs chugging by. A burly policeman stopped me and said that I had to get in line to get a taxi. It turned out that there must have been 50 people in line waiting for a taxi and it took at least 35 minutes. We started talking to the man in front of us; he said that he drives an hour each morning from his home in New Jersey to the train, rides the train for an hour to Manhattan, and then it takes another hour waiting for a cab and the ride to his office (and he reversed the procedure at night). He said the pay was good to make it worthwhile (he must have been paid very well to be willing to do this every day), but he still couldn't afford to live in Manhattan.

We left our bags at the hotel and then went to the Gray Line Station where we purchased a 2-day hop-on hop-off bus tour. The first day was the Downtown loop. We passed Times Square with a zillion flashing electronic billboards (picture what it must be like New Year's Eve!). We didn't get off until the 5th stop, which was the Empire State Building where we bought a ticket and took the elevator to the 86th floor (King Kong wasn't hanging around that day). We could see the borough of Queens to the east, Manhattan and the World Trade Towers to the south, and a green space (Central Park) to the north, with massive skyscrapers piercing the skyline all around us.

After going past a few more stops, we got off at Greenwich Village where we had a pastrami on rye sandwich at a classic New York deli—it was fascinating watching these New Yorkers rush in, rattle off a few words and have a massive sandwich suddenly appear. We expected to see more artists and free spirit types, but maybe it was too early for them. Most of the people didn't look any odder than us.

Going back to one of the designated stops, we hitched the next bus which took us down "Broadway," past City Hall, and to Battery Park, which is where we could see the famous lady (the Statue of Liberty) across the water in the distance. The next bus took us past the United Nations with its dark windows hiding international intrigue and diplomats; all we could see were the limousines lined in front of the buildings. We were taken past the famous Brooklyn Bridge (we sort of hoped that someone would try to sell it to us), and then past one of the Trump Towers where apartments were on sale ranging in price from $800,000 to $11 Million (now we knew why people lived three hours away in New Jersey).

After getting off the bus for the last time that day we walked to our hotel, and after a 30 second rest we met a friend (who was working in Manhattan) for dinner just off Broadway. After dinner we walked a dozen or so blocks with him to Central Park; we would never have done this by ourselves; we expected muggers and guns since it was getting dark.

Painting Pictures of the Big Apple

Instead we saw joggers and even families pushing baby carriages and people having a good time, and the experience was exhilarating.

You think that Vancouver and Toronto have traffic? Well multiply this by several factors and you have Manhattan. Orange lights don't mean the same thing as in other cities—cars and trucks don't stop until the intersection is clogged. The streets were jammed with pedestrians, and everyone jaywalks even between moving vehicles and all the cacophony and traffic. Everybody who isn't stuck in traffic in cars walks in New York City to and from the subway.

The second morning we trudged back to the Gray Line Station and took the Uptown Loop, starting past the prestigious Columbia University, the 'poor mans' CCNY (City College of New York), the impressive Lincoln Center, the Dakota Apartments (where John Lennon used to live; we craned our necks to see if Yoko Ono was standing at her window to wave to us). We saw Riverside Church, which the tour guide told us had more millionaires than there were in heaven, which I thought was rather a weird statement but typical NYC hyperbole. We drove through Harlem, which was fascinating, and saw some poor areas and the "Salvation and Deliverance Baptist Church" which is a powerful name for a church (no mention was made of the number of millionaires). The only place we hopped off was at the New York Metropolitan Museum where we spent a couple of hours. Then we hopped back on the bus and went past the Rockefeller Center and the historic Radio City Music Hall.

After a late lunch we walked down Fifth Avenue. My wife said she would have to go shopping for new clothes and accessories before she could shop at the extra fancy stores. But we did walk into St. Patrick's Cathedral. I'm not sure what event was happening, but there was a humongous processional that had just started. Hundreds of church officials in white robes with black fuzzy hats marched 3x3 down one side aisle and back up the center aisle to the front. This is on a Friday afternoon. Then men with white robes and no hats marched 3x3 coming down the other side aisle and up the center, then women in black nun uniforms came 3x3, then more men in colourful robes 3x3, then more men 3x3, some with no hats and some with white pointed hats. Overlaying this impressive display was a lady with the most beautiful voice singing, and the music from the great pipe organ was awesome. The church and the music made all of this very worshipful.

One reason New York is expensive is that everyone expects to be tipped. In the course of our two day bus tour we had four or five tour guides as we hopped on or off and they all expected a tip. When we stored our bags for the day, tip, when we picked them up, tip, when we took a taxi to the train station, tip. But I'll give you a tip—visit New York City! And we didn't even visit the boroughs of Brooklyn or Staten Island or Queens.

VIGNETTE 96

Snowbirds

THERE ARE VARIOUS kinds of snowbirds and when you hear this term you might think about Canada's military aerobatics or air show flight demonstration team that visits airshows across the country and around the world to demonstrate the skill, professionalism, and teamwork of Canadian Forces personnel. But I'm thinking of the kind where Canadians flee the terrors and rigors of the cold Canadian winter to find sun and solace in the southern U.S.

Most of these snowbirds are age 55 or older and have the resources to spend a few months in a warmer climate. Health care is a significant issue for many, and the rule generally is that a person cannot be outside of Canada for more than 6 months for their health insurance to be valid. Even then, the cost of health insurance may be prohibitive for those of modest means and who have "pre-existing conditions" that drive up the cost of insurance.

There are other complications if Canadians stay in the U.S. for 183 days or more in a calendar year, since you are then considered a "resident alien" for U.S. tax purposes and must file a regular U.S. tax return. Even if snowbirds regularly stay in the U.S. for up to 6 months each year, they could meet what is called the "substantial presence" test and be subject to U.S. tax. Uncle Sam is in dire financial straits these days and will look for tax revenue wherever possible.

There even is a Canadian Snowbird Association, which provides information to prospective snowbirds on issues ranging from health insurance (recommending everyone have a "personal health record" to facilitate proper care by an unfamiliar health care professional"), to finding places to rent (or buy), to finding campgrounds for those with recreation vehicles, to describing local attractions. They even have "Lifestyle Presentations" for those who feel they need assistance in making sure that they get the most out of their southern sojourn.

There are various kinds of snowbirds, including Florida snowbirds primarily from Ontario and Quebec, Texas snowbirds, Arizona snowbirds and California snowbirds. Some American citizens may resent the Canadian invasion each fall, but when they stop to think about all the money that Canadians are spending they tend to relax and make friends, and even some money. The state of Florida, for example, does lots of advertising to make Canadians aware of what's available and they provide lots of information, like where to find approved accommodation, what deals are offered for golfers, and local advice such as "do not feed the alligators." Other southern states also advertise consistently to attract and inform potential snowbirds, since they often constitute a substantial portion of revenue for the local and state governments.

Since many snowbirds leave the rapidly cooling Canadian climate in October or even November, they face the dilemma of wanting to be back home in Canada for Christmas. Some are fortunate enough to be able to afford to fly home for 10 days or so or to bring their family south for a short time, while others decide to fly home for the Christmas break. This return to Canada may enable them to stay in the U.S. for a longer period of time.

Some "mini snowbirds" may go south for only 2 months or so, either for financial considerations or because they don't want to be real snowbirds and be away from home for a longer period of time. This describes the approach that we have taken a couple of times. In these cases there is the question of when to go and what type of accommodation may be available, so the October to November or early December time period has some advantages. The "high season" down south is January to March, so costs are higher during this period. Many rental accommodations then want lease agreements for the whole season and aren't interested in a one or two month stay. The same may apply to those with RV's since many RV parks want folks to stay longer, and those parks that provide short term spaces are generally full. While both RV parks and rental accommodations are more available and less expensive before Christmas, the disadvantage of going south from late September to early December is that the coldest months of the Canadian winter are still ahead, but at least we are able to enjoy every 85 degree sunny day while down south. Driving back home from Arizona or southern California can be a challenge when going through the high mountain passes in northern California and Oregon, with the distinct possibility of snow, so many choose the coastal route instead.

VIGNETTE 97

Wonders of the World

IN HIGH SCHOOL Social Studies I recall hearing about the mysterious and exotic and impressive "Seven Wonders of the World." This was a list of perhaps the most remarkable man-made creations in the world. When I recently investigated this further it turned out that the original list was based on Greek guide books extolling only works located around the Mediterranean. The number seven was chosen because the Greeks believed it to be the representation of perfection.

The seven wonders included the Great Pyramid of Giza, the Hanging Gardens of Babylon, the Statue of Zeus at Olympia, the Temple of Artemis at Ephesus, the Mausoleum of Maussollos at Halicarnassus, the Colossus of Rhodes, and the Lighthouse of Alexandria.

The only ancient world wonder that still exists is the Great Pyramid of Giza, built by the Egyptian Pharaoh Khufu about 2560 BC to serve as a tomb when he died. It is possible that the Hanging Gardens of Babylon might have never existed except in Greek poets and historians' imagination, but legend has it that Nebuchadnezzar II (604–562 BC) built the Hanging Gardens to please his wife. The massive and impressive Statue of Zeus at Olympia was built about 350 BC to recognize the god in whose honor the Ancient Olympic Games were held. It was destroyed by fire in 462 AD. The Temple of Artemis (Diana) at Ephesus was built in honor of the Greek goddess of hunting and wild nature and was described by many visitors as not just a temple but as the most beautiful structure on earth.

Wonders of the World

The Mausoleum at Halicarnassus (in a Roman province in Asia) is the burial place of the ancient king Maussollos who died about 353 BC; this is a huge beautiful rectangular structure about 140 feet high with decorations and statues that adorned the outside at different levels. Rhodes was a Greek city state and they built the giant colossus statue of their sun god Helios as a symbol of unity and victory over a rival city state. Unfortunately it was destroyed by an earthquake 56 years later, about 226 BC. The Lighthouse in the Egyptian city of Alexandria was the only practical structure of the seven wonders, since it helped ensure a safe return for sailors to the harbour. It was the tallest building on earth for many years and the great mirror's reflection could be seen more than 35 miles off-shore.

People love to make lists, so of course we now have a list of the Seven Wonders of the middle ages, or the Medieval World. The list usually includes Stonehenge, the Roman Colosseum, the Catacombs of Kom el Shoqafa, the Great Wall of China, the Porcelain Tower of Nanjing, the Hagia Sophia and the Leaning Tower of Pisa. An expanded list would add the Taj Mahal and the Cairo Citadel.

Stonehenge in southern England is composed of earthworks surrounding a circular setting of large standing stones erected around 2500 BC. There has been much speculation how the huge stones could have been hauled there and erected in such precise fashion. A person standing within Stonehenge who faces northeast through the entrance towards the "heel stone" will see the sun rise above the stone at summer solstice. The Colosseum in Rome was completed in 80 AD. Its 50,000 spectators watched gladiatorial contests, the killing of Christians by wild animals, and public spectacles such as mock sea battles, animal hunts, executions, and re-enactments of famous battles. It is still one of Rome's most popular tourist attractions. The Catacombs of Kom el Shoqafa (meaning mound of shards) is a historical archaeological site located in Alexandria, Egypt. It consists of a series of Alexandrian tombs, statues and archaeological objects and was used as a burial chamber from the 2nd century to the 4th century.

The Great Wall of China is a series of stone and earth fortifications in northern China, built originally to protect the northern borders of China against intrusions by various nomadic groups. The wall has been rebuilt and maintained from the 5th century BC through the 16th century. The Great Wall stretches over 5,500 miles from east to west, along an arc that roughly delineates the southern edge of Inner Mongolia. This is made up of about 3,900 miles of actual wall (much of which has now deteriorated), 223 miles of trenches and 1,387.2 miles of natural barriers such as hills and rivers. The Porcelain Tower of Nanjing was a pagoda constructed in the 15th century on the Yangtze river in Nanjing, China. It was one of the largest buildings in China at a height of 260 feet, with nine stories and a staircase in the middle, and the top of the roof was marked by a golden pineapple. The Hagia Sophia is a former Orthodox basilica, then a mosque, and now a museum in Istanbul, Turkey.

The Leaning Tower of Pisa, situated behind the Cathedral, is a freestanding bell tower in the Italian city of Pisa. The tower now leans at about 4 degrees, meaning that the top of the tower is displaced horizontally 3.9 meters) from where it would be if the structure were perfectly vertical. The tower began to sink after construction had started on the second floor in 1178 due to an inadequate foundation set in weak, unstable soil, but construction periodically continued and the seventh floor was completed in 1319. Authorities spent millions of Euros between 1990 and 2001 to structurally strengthen the base and prevent eventual collapse. The Taj Mahal in Agra, India was built by an Emperor in memory of his wife; it has been called the jewel of Muslim art in India and a masterpiece of the world's heritage, plus the unofficial symbol of India. The Cairo Citadel is a massive Islamic medieval fortress built about 183 AD that still exists today.

People can build all kinds of wonderful things, but it's difficult to top Mother Nature and God, the Creator. One list of the "Seven Natural Wonders of the World" is the Aurora Borealis (the Northern Lights), the Grand Canyon in Arizona, Paricutin (a cinder cone volcano in Mexico), Victoria Falls (the widest and highest waterfalls in the world, on the border of Zambia and Zimbabwe), the Great Barrier Reef in Australia (the largest coral reef system in the world, stretching over 2600 km), Mount Everest (the highest mountain in the world at 29,029 feet or 8,848 km, in the Himalaya Mountains between Nepal and Tibet), and the Harbor of Rio de Janeiro (the largest bay in the world, and surrounded by unique mountains). Well, that's the top seven, and it's a pretty impressive list. But once you start on lists like this it's hard to stop. For example, there are many other great natural wonders, including the Sahara Desert, Mount Fuji, the Dead Sea, Mount Etna, Rock of Gibraltar, Redwood National Forest, Everglades National Park, Ayers Rock, and the Amazon Rain Forest. I wouldn't mind adding the Columbia Icefields and Maligne Lake, both near Jasper, Alberta.

We have ancient wonders and medieval wonders, and of course we now have modern wonders of the world. These include the Channel Tunnel between England and France, the CN Tower in Toronto, which used to be the tallest structure in the world, the Empire State Building in New York City, the Golden Gate Bridge in San Francisco, the Itaipu Dam which is the world's largest hydroelectric project that produces nearly 25 percent of Brazil's supply, the Zuiderzee Works, which are a man-made system of dams and water drainage works by damming of the Zuiderzee, (a large inlet of the North Sea) to improve flood protection and create additional land for agriculture, and the Panama Canal which transports over 14,000 ships a year through its 77 km ship canal joining the Atlantic Ocean and the Pacific Ocean.

New wonders are popping up all the time. The world's tallest building in the world as of 2010 was the 828 m (2,717 ft) Burj Khalifa in Dubai, United Arab Emirates, but by the time you read this there may be more massive and taller buildings in places like Dubai.

VIGNETTE 98

Canada's Wonders

AFTER I CHECKED out the great wonders of the rest of the world, I wondered (!) how Canada had approached this issue. I assumed that we didn't have any thousand-year-old man-made wonders, but I knew that we had many natural wonders. Well, it turns out that our dear Canadian icon, the CBC, has asked this question. In 2007 CBC television and radio sponsored a competition called "The Seven Wonders of Canada." They sought to determine these "wonders" by receiving nominations from viewers, and then from on-line voting of the short list.

I'm a semi-regular listener of CBC radio, but I don't remember this survey, and the very first "wonder" on the list is a bit of a head scratcher and leads me to believe that, like many other things in Canada, the exercise was dominated by Ontario. The first item was the Sleeping Giant Provincial Park near Thunder Bay; this park is indeed apparently very beautiful, but have you ever heard of it? The next wonder listed was Niagara Falls, which indubitably is spectacular and worthy of top billing. Next was the Bay of Fundy and its massive tides; I have seen the effects of these 40 foot tides at the Hopewell Rocks near Moncton, New Brunswick. Unfortunately, few Canadians are able to visit Nahanni National Park, but it gives one a dramatic feel for the power and beauty of nature untrammeled by people. As a kid growing up on a farm in Alberta I can remember seeing the Northern Lights in winter, and they are truly one of the greatest natural shows you will ever see. Brave souls from many countries sojourn to places like Whitehorse to see the rainbows of moving lights, perhaps because there is less light pollution in less populated areas. The Canadian Rockies also made the top list. They are not the highest mountains in the world, but they are our mountains and they inspire feelings of the power of nature. The Cabot Trail was also on the list, and this winding road provides a multitude of vistas as you wait to see what might be around the next corner.

Since the Sleeping Giant is a question mark for me, I'm adding four more wonders to make the top 10. This is tough, because there are so many choices and my bias leans towards places I have seen or have heard lots about. I like the 13 km Confederation Bridge which overcame the powers of winter ice flows and island resistance to link Prince Edward Island and New Brunswick. The Rideau Canal is a 202 km World heritage site stretching from Kingston to Ottawa, linking with various lakes and rivers along the way. It was originally built for defensive purposes to enhance out protection from the Americans, was also a commercial waterway, and is now a boating aficionado paradise. I've never been to Haida Gwaii (previously known as the Queen Charlotte Islands), but this area reminds one of the role of the Haida people and their ability to preserve and enhance our heritage. There are two main islands and about 150 smaller islands off the coast of BC. I'm including Long Beach in BC and the "Singing Sands Beaches of PEI (said to be one of the most intriguing sites in Canada) as a double entry on this list. Who would have thought that a cold northern country could have such beautiful beaches?

Well, that's the top 10, but we are far from listing all the wonders of Canada. There are some nominations that may surprise you. They are rather odd at first blush as wonders, but are distinctly and emphatically Canadian. One of the nominations was the canoe. This seems very odd since many other cultures relate to the canoe much more than Canadians, even with our early exploration as well as modern day sightseeing. Two much more understandable items are the Grey Cup, recognizing Canadian Football League supremacy, and hockey's Stanley Cup which is the oldest trophy in North America and said to be the most difficult sports trophy to win. The Inuit Igloo fits in this section; snow was used to insulate their whalebone and hides houses since it is a good insulator. While outside temperatures may be as low as −45 °C, inside the temperature is comfortably warm. The giant Vegreville Egg featuring Ukrainian designs was on this list, and can serve as a representative for many other Canadian communities (like Abbotsford's giant raspberry or Prince George's anthropomorphic Mr. PG, or Shediac's "largest in the world" lobster, or go to Nunavut to see Rankin Inlet's Inuksuk), which have adopted distinctive icons as road side attractions to describe an important feature of their city. Another typical Canadian entry is the Montreal bagel, which differs from the prototype New York bagel by being smaller and sweeter. The Hartland Covered Bridge at almost 1300 feet is the world's longest covered bridge, and crosses the St. John River in New Brunswick. Drumheller, Alberta is home to Dinosaur Valley and the haunting badlands.

I can't ignore Ontario, so I must add Ottawa's Library of Parliament, which is fascinating and hopefully useful to our parliamentarians. But if we're talking natural wonders I don't see how one can avoid mentioning Peggy's Cove in Nova Scotia or Percé Rock in the Gaspe Peninsula, which is a major Quebec icon, and the dramatic Gros Morne National Park on the west coast of Newfoundland. The Saguenay Fjord at over a hundred kilometres is one of the largest, yet least known fjords in the world. It allows the waters from the Atlantic Ocean and the Gulf of Saint-Lawrence to flow into the heart of the Quebec's region. Saskatchewan has a couple of modest nominations, namely the "Crooked Trees," a grove of genetically mutated aspens, and the Cypress Hills, the edge of the Rocky Mountains, are the highest point in that province. BC has its own Cathedral Grove on Vancouver Island, Canada; the park protects the old growth west coast rainforest. Manitoba has thousands of lakes, and Grand Beach on the eastern shore of Lake Winnipeg is a grand place. I once took my aging parents to Wreck Beach near the University of BC but they didn't notice anything wonderful.

The Dempster Highway connects the Klondike Highway near Dawson City in the Yukon for 736 km to Inuvik in the Northwest Territories. During the winter months, the highway extends another 194 km to Tuktoyaktuk, on the northern coast of Canada, using frozen portions of the Mackenzie River delta as an ice road where men and women battle brutal weather conditions in isolated areas. The highway provides some of the most impressive vistas that you will see in Canada. I had never heard of the Tuktoyaktuk Pingos, which are snow cored conical mounds and hills in the flat slumbering north of Canada, but they are a natural wonder.

Well, there are many more wonders to see in Canada but we can't list them all. One last remarkable and poignant Canadian wonder isn't even in Canada—this is the Vimy Memorial in France, dedicated to the Canadian soldiers who died in WWI.

VIGNETTE 99

The Ultimate Trivia Quiz

MY BRAIN STORES all sorts of trivial information, much of which is eminently un-useful and impractical. For example, even when I periodically watch "Jeopardy" I do spectacularly poorly. If you can answer most of the following questions, it doesn't necessarily mean that you are brilliant; it may simply indicate that your brain is as unusual as mine. And if you do poorly, that may show that you reserve your brain function to process more useful information.

1. Who is the actor that is known for playing Mr. Bean?
2. When he was still playing hockey in the NHL, where was Wayne Gretzky's "office"?
3. What country is called the world's largest democracy?
4. What was the primary invention of Alfred Nobel, the Nobel Prize initiator?
5. Who founded the Apple Computer, Inc.?
6. What Canadian Prime Minister won the Nobel Peace Prize?
7. Name the lead actor in the most famous well-loved version of the Christmas Carol movie.
8. The assassination of what royal person led directly to the start of WWI?
9. Who painted the Mona Lisa?

10. What US founding father signed the US Declaration of Independence in large script and said he hoped that King George could read his flourishing signature?
11. What is the largest church building in the world?
12. Name the 2011 inductee to the "27 Club".
13. Who was the pilot of the Spirit of St. Louis first crossing the Atlantic?
14. What pro football quarterback said that he wasn't a genius; that guys like "Norman Einstein" were the geniuses.
15. Fill in the blank to complete the name of this country: Trinidad and _____.
16. What U.S. President said "Ich bin eine Berliner"?
17. Name the Great Lakes in order starting from the west.
18. Who was the Norwegian Prime Minister whose name became synonymous with "traitor"?
19. Who wrote the Lord of the Rings stories about Hobbits?
20. What famous Queen came to see if King Solomon really was wise?
21. What famous Queen became romantically involved with Marc Antony?
22. What famous Greek is said to have run out on the street naked after taking a bath shouting "Eureka!" after making a great discovery?
23. Who is generally credited with inventing the light bulb?
24. Who painted the ceiling of the Sistine Chapel?
25. Who was the British Prime Minister who appeased Hitler before and at the start of WWII?
26. In what year was the King James Bible published?
27. Who was the primary leader of the Protestant Reformation?
28. What Ukrainian city was the site of a major nuclear power plant melt down?
29. What singer was known as "the man in black"?
30. What do Barry Bonds, Mark McGwire and Sammy Sosa all have in common?
31. What man is said to have killed the most people in all of history?
32. What is the smallest country in the world?

100 Vignettes

33. What is the fastest animal in the world over short distances?
34. Who was the U.S. Teamster's Union leader who was killed but his body never found?
35. What was the hometown of The Beetles?
36. What was the name of Elvis Presley's home in Memphis, Tennessee?
37. What is the most difficult play in baseball?
38. What was Albert Einstein's first (boring) job?
39. From what North American city did Marconi send the first trans-Atlantic cable?
40. Name the city in which the Rose Bowl is played each January.
41. Who was the first freely elected President of Russia?
42. Name the Russian mystic charlatan who gained the trust of Tsarina Alexandra because he said he could treat her son's hemophilia.
43. Who was the Italian Fascist dictator in Italy during WWII?
44. Name the female aviator who disappeared while trying to fly solo around the world in 1937?
45. What is regarded as the southern tip of South America?
46. What Canadian company made some of the first snowmobiles?
47. List the colors of the rainbow in order starting with violet.
48. What is the nickname of New York City?
49. What is the last book of the Bible?
50. Who invented the printing press in 1439?

1) Rowan Atkinson; 2) Behind the opposition's goal; 3) India; 4) Dynamite; 5) Steve Jobs (and Steve Wozniak); 6) Lester B. Pearson; 7) Alistair Sim; 8) Archduke Ferdinand of Austria; 9) Leonardo da Vinci; 10) John Hancock; 11) St. Peter's Basilica in Rome; 12) Amy Winehouse; 13) Charles Lindbergh; 14) Joe Theismann; 15) Tobago; 16) J.F. Kennedy; 17) Superior, Michigan, Huron, Erie, Ontario; 18) Quisling; 19) J.R.R. Tolkien; 20) Queen of Sheba; 21) Cleopatra; 22) Archimedes; 23) Thomas Edison; 24) Michelangelo; 25) Neville Chamberlain; 26) 1611; 27) Martin Luther; 28) Chernobyl; 29) Johnny Cash; 30) baseball players who hit home runs who used steroids; 31) Joseph Stalin; 32) Vatican City; 33) the cheetah is the fastest land animal (peregrine falcons

are the fastest in the air); 34) Jimmy Hoffa; 35) Liverpool, England; 36) Graceland; 37) the unassisted triple play; 38) a Swiss Patent Clerk; 39) St. John's, NL; 40) Pasadena, California; 41) Boris Yeltsin; 42) Rasputin; 43) Benito Mussolini; 44) Amelia Earhart; 45) Cape Horn; 46) Bombardier; 47) Violet, Blue, Green, Yellow, Orange, Red; 48) The Big Apple; 49) Revelation; 50) Gutenberg.

VIGNETTE 100

And Now for Something Completely Different— Harry and the Leeches

HARRY KELSEY APPARENTLY was just your ordinary everyday guy. He had a boring job in the shipping department of a small manufacturing company even though he had two college degrees in ancient history and one science degree for dummies. But if you dug a little deeper, Harry was no dummy. Harry liked to go to the beach to collect shells on his days off; sometimes, if it was nice, he would go in the evenings after work. Harry's girlfriend didn't like the water, so she rarely went along with him, and even if she did, Sally wouldn't go into the water. Harry's other hobby was amateur medicine, since he harboured a secret desire—which he wouldn't admit even to himself—to be a Paramedic so he could speed down the streets with the siren blaring and people showing him the respect that he never saw in his current job.

So it was that Harry knew he was in deep, deep trouble one day as he waded out of the water and saw a 17 inch European medical leech, *Hirudo medicinalis*, attached to one leg and several dozen slightly smaller, but still massive, ones firmly implanted on his other leg. Harry knew that these rascals have been used for clinical bloodletting for thousands of years under controlled circumstances, and Harry instinctively knew that his own circumstances were rapidly spinning out of control. Of course, Sally, who moments earlier had been demurely relaxing on a lawn chair reading the May issue of The National Enquirer, rushed screaming to Harry's dilapidated sedan with a beach towel over her head when she saw those gruesome wiggling things on Harry's legs.

And Now for Something Completely Different—Harry and the Leeches

Harry guessed that the obnoxious, ugly creatures (let's call them Larry and friends) attached to his legs were maybe 50 times bigger than average and he could already feel himself getting weak. He could hardly bear to look at the "suckers" that the leeches use to connect to a host (his legs in this case) for feeding. He hadn't noticed his unwelcome friends earlier because of the anesthetic that leeches use to prevent the host from feeling their piggy back approach search for red liquid food. Harry was concerned about the combination of mucus and suction used by Larry and friends to stay attached and secrete an anti-clotting enzyme, hirudin, into his blood stream. He panicked even more, if that were possible, when he wondered if Larry and friends were from the family Gnathobdela which he knew to be "jawed" leeches armed with teeth. Whoever, or whatever, Larry and his friends were, they weren't coming off Harry's legs when he tried to brush them off. They would leave on their own only after having their fill.

He rushed to his car screaming "Call 911! Call 911!" but as Sally furiously scrambled to obey, she yelled that her cell phone was dead. Harry decided that he would need to drive himself to the hospital, but as he reached for the door he heard an ominous click. "There's no way that you're bringing those yucky things into the car anywhere near me" Sally yelled. "I'll drive to a store, find a phone and get help". She climbed and slithered over to the driver's seat and took off. For the 7th time in his life, Harry cried while he reviewed his Greek and Roman and ancient Indian history, forgetting that removing Larry and friends should be his top priority (But Harry was never good at setting priorities; would this prove to be his downfall?).

He recalled that use of leeches in medicine dates as far back as 2,500 years ago when they were used for bloodletting in ancient India, since leech therapy was explained in ancient Ayurvedic texts. He remembered his ancient Greek history where bloodletting was practiced according to the humoral theory, which proposed that when the four humors, blood, phlegm, black and yellow bile in the human body were in balance, good health was guaranteed. Even the 5th century B.C. philosopher/doctor Hippocrates B.C. used bloodletting leeches to balance the humors and to rid the body of the plethora. Even though the use of leeches in modern medicine made its comeback in the 1980s Harry didn't have much confidence that the local doctor, who was really a veterinarian, would be able to help. The "Doc" didn't know that one of Harry's friends had his ear cut off in an accident and they used leeches to stimulate blood flow to help the sewed-back-on ear connection heal. He hadn't been told that in reconstructive surgeries one problem that arises is venous congestion due to inefficient drainage leading to blood clots, which is prevented by the anticoagulant (hirudin) in the leeches' saliva, leading to healing.

Well, let's cut right to the end of the story. When Doc Hammer showed up in his beaten old half ton (Sally eventually did find a telephone), all he saw was about 30 giant leeches lying on the sand, gorged with blood. And Harry, why he hadn't even tried to use his fingernails to remove the leeches, no one will ever know. All that remained of Harry was what looked something like some bones under a big balloon after almost all the air is sucked out, with some hair on top.

WinePressPublishing
Great Books, Defined.

To order additional copies of this book call:
1-877-421-READ (7323)
or please visit our website at
www.WinePressbooks.com

If you enjoyed this quality custom-published book,
drop by our website for more books and information.

www.winepresspublishing.com
"Your partner in custom publishing."

CPSIA information can be obtained at www.ICGtesting.com
Printed in the USA
LVOW13s0929051213

363899LV00005B/47/P